INVARIANT THEORY, Old and New

INVARIANT THEORY, Old and New

JEAN A. DIEUDONNÉ

Faculty of Science, Mathematics
University of Nice
Parc Valrose
Nice, France

JAMES B. CARRELL

Division of Mathematical Sciences
Purdue University
Lafayette, Indiana

1 9 7 1

ACADEMIC PRESS New York and London

ACADEMIC PRESS, INC.
111 Fifth Avenue, New York, New York 10003

United Kingdom Edition published by
ACADEMIC PRESS, INC. (LONDON) LTD.
Berkeley Square House, London W1X 6BA

LIBRARY OF CONGRESS CATALOG CARD NUMBER: 79-153663
AMS (MOS) 1970 SUBJECT CLASSIFICATIONS: 15-03, 15A72, 20G05

Reprinted from *Advances in Mathematics*, Volume 4, Number 1, February 1970

PRINTED IN THE UNITED STATES OF AMERICA

Contents

Preface

Invariant theory has already been pronounced dead several times, and like the phoenix it has been again and again rising from its ashes. The first period in the history of the theory culminated with the discovery of the so-called "symbolic method" which in theory allowed the computation of all invariants by a quasi-mechanical process. But it was soon realized that, except in a very few simple cases, the actual computation would lead to enormous labor, disproportionate with the interest of the outcome, especially in a period when all calculations were done by hand (it might be worthwhile to push the XIXth Century computations of invariants a little further along, with the help of modern computers). Partly for that reason, the next problem in the theory was the search for "fundamental systems" of invariants, i.e., finite sets such that any invariant would be a polynomial in the fundamental invariants. It is well known that the existence of such systems was proved by Hilbert in 1890, in a brilliant paper which made him famous overnight and which may be considered as the first paper in "modern algebra," by its conceptual approach and methods. But Hilbert's success also spelled the doom of XIXth Century invariant theory, which was left with no big problems to solve and soon faded into oblivion.

The first revival was prompted by the developments (I. Schur, H. Weyl, E. Cartan) of the global theory of semi-simple groups and their representations around 1935, when it was realized that classical invariant theory was really a special case of that new theory; this was clearly shown in H. Weyl's famous book "Classical Groups," but again a lack of outstanding problems was probably the reason why important new developments failed to materialize after the publication of that book. Only very recently have new stirrings of life been perceptible again; this is mainly due to the work of D. Mumford, who realized that invariant theory provided him with some of the tools he needed for his solution of the problem of "moduli" of algebraic curves. His new approach to the theory has been to subsume it under the more general question of defining "spaces of orbits" (with suitable structures) of algebraic groups acting on algebraic varieties; in so doing, he discovered that some essential techniques and ideas pertaining to such questions lay buried in a beautiful and long forgotten paper which Hilbert had published in 1893. In his book on "Geometric Invariant Theory," Mumford has modernized and greatly generalized these ideas, using the language of the theory of schemes, as well as important contributions of Chevalley, Nagata, Iwahori, Tate, Tits, and himself; it seems quite likely that this book should exert a deep influence in the years to come and that its methods should prove useful in many other problems.

In these lectures (given at the University of Washington in 1967), I have tried to provide an elementary introduction to invariant theory; more systematically than in Weyl's book, I have tried to describe it as part of the theory of linear representations of groups, without neglecting to link it to its geometric origin. The first two chapters are essentially a description of the "symbolic method" understood in that manner; the finiteness theorem and the Nagata counter-example to its extension to all algebraic groups form the subject of Chapter 3, and Chapter 4 is to be considered as an introduction to the Mumford theory.

I have tried throughout to make the book accessible to readers having only a bare knowledge of elementary algebra at the undergraduate level; the necessary prerequisites of noncommutative algebra have been developed in an Appendix. Only in Chapters 3 and 4 do I need some less elementary results from commutative algebra; they are all to be found in the Zariski–Samuel treatise on the subject.

The book could not have taken its present form without the active collaboration of Professor J. B. Carrell, who has taken great pains to write in readable form many arguments which had remained very sketchy in the oral presentation of the lectures. I am happy to thank him most heartily for his invaluable help.

Nice, December 1968

J. Dieudonné

Chapter 1. Elements of the Theory of Invariants

1. The Notion of an Invariant

Let Γ be a group acting on a set E; that is, suppose there exists a map $(\sigma, x) \to \sigma \cdot x : \Gamma \times E \to E$ with the properties:

(1) $(\sigma \cdot \tau) \cdot x = \sigma \cdot (\tau \cdot x)$ for all $x \in E$, $\sigma, \tau \in \Gamma$, and

(2) $\epsilon \cdot x = x$ for all $x \in E$, where ϵ is the identity of Γ.

Thus the map $\mu_\sigma : E \to E$ defined by $x \to \sigma \cdot x$ is a bijection of E, for by (1) and (2), $\mu_{\sigma^{-1}}\mu_\sigma = \mu_\sigma\mu_{\sigma^{-1}}$ is the identity map of E. By (1) the mapping $\sigma \to \mu_\sigma : \Gamma \to \mathscr{G}_E$, the group of bijections of E, is a homomorphism.

Definition. An element $x \in E$ is *Γ-invariant* or, simply, *invariant*, if $\mu_\sigma(x) = x$ for all $\sigma \in \Gamma$. A subset F of E is *Γ-stable* or, simply, *stable*, if $\mu_\sigma(F) \subset F$ for all $\sigma \in \Gamma$.

Associated actions of Γ

(1) If $(E_\alpha)_{\alpha \in I}$ is a family of sets on which Γ acts, then Γ acts on $\mathscr{E} = \prod_{\alpha \in I} E_\alpha$ by $\sigma \cdot (x_\alpha) = (\sigma \cdot x_\alpha)$ for $\sigma \in \Gamma$.

(2) Γ acts on $\mathscr{P}(E)$, the power set of E, by $\sigma \cdot F = \sigma(F)$. A subset of E which is invariant under this action of Γ is stable with respect to the action of Γ on E, but not conversely.

(3) Suppose Γ acts on the sets E and F. Let $\mathscr{F}(E, F) = F^E$ be the set of all mappings $E \to F$. Then Γ acts on $\mathscr{F}(E, F)$ in a natural way. In fact, if we imbed $\mathscr{F}(E, F)$ in $\mathscr{P}(E \times F)$ by identifying a function with its graph and use the actions defined in (1) and (2), we see that for $u \in \mathscr{F}(E, F)$, $\sigma \cdot u : \sigma \cdot x \to \sigma \cdot u(x)$ is the natural action. Equivalently, for $u \in \mathscr{F}(E, F)$ and $\sigma \in \Gamma$, we define $\sigma \cdot u \in \mathscr{F}(E, F)$ by $(\sigma \cdot u)(x) = \sigma \cdot u(\sigma^{-1} \cdot x)$.

Definition. $u \in \mathscr{F}(E, F)$ is a *concomitant* of Γ if u is invariant under the action of Γ on $\mathscr{F}(E, F)$. Thus u is a concomitant if for all $x \in E$ and $\sigma \in \Gamma$, $\sigma \cdot u(x) = u(\sigma \cdot x)$.

Definition. Suppose Γ acts on each set in a family $(E_\alpha)_{\alpha \in I}$ and on F. We say that $u \in \mathscr{F}(\mathscr{E}, F)$ is a *simultaneous concomitant* of Γ if $\sigma \cdot (u(x_\alpha)) = u((\sigma \cdot x_\alpha))$ for each $\sigma \in \Gamma$.

Remark. In the literature the terms covariant and equivariant are sometimes used for concomitant. As these terms have many other usages, we shall, for the time being, use concomitant exclusively.

Now let E be a vector space over a field K and suppose Γ acts *linearly* on E; that is, for each $\sigma \in \Gamma$, $\mu_\sigma : E \to E$ is linear. Thus μ_σ is an element of $\mathbf{GL}(E)$, and the mapping $\sigma \to \mu_\sigma : \Gamma \to \mathbf{GL}(E)$ is a *linear representation* of Γ.

Definition. A homomorphism $\chi : \Gamma \to K^* = K - \{0\}$ is called a *character*, or *abelian character*, of Γ in K.

Definition. A *relative invariant* of Γ of weight χ is a mapping $u : E \to K$ such that for all $\sigma \in \Gamma$ and $x \in E$, $u(\sigma \cdot x) = \chi(\sigma) \cdot u(x)$. A relative invariant of weight $\chi = 1$ is called an *absolute invariant* of Γ.

Thus a relative invariant of Γ of weight χ on E is a concomitant $u \in \mathscr{F}(E; K)$ for the action of Γ on K given by χ.

Examples. (1) Let Γ be the group of isometries of $\mathbf{R}^2$. Γ is isomorphic to the linear group of transformations of $\mathbf{R}^3$ of the form

$$\begin{pmatrix} & & 0 \\ & \underline{U} & 0 \\ a & b & 1 \end{pmatrix}$$

for $\underline{U} \in \mathbf{O}(2)$ and a, b real numbers. Consider $\mathbf{R}^2$ as the plane in $\mathbf{R}^3$ defined by $\xi_3 = 1$. For $p, q, r \in \mathbf{R}^2$, $(x - p \mid q - r) = 0$ represents the equation of the perpendicular to $\vec{qr}$ through p. The linear form $(x - p \mid q - r)$ is a concomitant of Γ in the dual space of $\mathbf{R}^3$.

(2) Let $E = \mathbf{R}^2$ and $\Gamma = \mathbf{GL}(2; \mathbf{R})$. For $x = (\xi_1, \xi_2)$ and $y = (\eta_1, \eta_2)$, let $[x, y] = \det(\begin{smallmatrix}\xi_1 \eta_1 \\ \xi_2 \eta_2\end{smallmatrix})$. A simple computation shows that $[\sigma x, \sigma y] = \det(\sigma)[x, y]$ so $[\ ,\ ]$ is a relative (simultaneous) invariant of weight $\chi(\sigma) = \det(\sigma)$.

(3) Let E be the space of bilinear forms on $\mathbf{R}^n$, and let $\Gamma = \mathbf{GL}(n, \mathbf{R})$ For $\underline{U} \in E$ set $\sigma \cdot \underline{U} = {}^t\sigma \cdot \underline{U} \cdot \sigma$. Then $\det(\sigma \cdot \underline{U}) = \det^2(\sigma) \cdot \det(\underline{U})$ so $\det(\underline{U})$ is a relative invariant of weight $\det^2(\sigma)$.

2. Rational Concomitants

From now on K is a field of *characteristic* 0. E and F will denote finite dimensional K-vector spaces, and Γ will denote a group acting *linearly* on E and F. Let $e_1, \ldots, e_m$ be a basis for E and $f_1, \ldots, f_n$ a basis for F.

Definition. A concomitant $u : E \to F$ is said to be a *rational concomitant* if $u(\Sigma_j \xi_j e_j) = \Sigma_k u_k(\xi_1, \ldots, \xi_m) f_k$ where each u_k is a rational function of $\xi_1, \ldots, \xi_m$. u is a *polynomial concomitant* if each u_k is a polynomial.

We remark that the notion of rational (resp. polynomial) concomitance is independent of the choice of the bases $e_1, \ldots, e_m$ of E and $f_1, \ldots, f_n$ of F.

For some types of groups Γ we shall eventually determine all rational concomitants on a finite dimensional vector space over an algebraically closed field K of characteristic 0. The class of rational concomitants on a finite dimensional vector space over a field of characteristic 0 is particularly pleasant since it is completely determined by the linear concomitants on the tensor powers of the space, as we shall see in the next four propositions.

Let us first recall some facts from algebra. For a more detailed discussion, the reader is referred to *Modern Algebra* by van der Waerden or any other elementary textbook on algebra.

Proposition (Weyl's principle of irrelevancy of algebraic inequalities). *Let I be an infinite integral domain with identity, and let $x_1, \ldots, x_n$ be indeterminates over I. Suppose that $f, g_1, \ldots, g_r \in I[x_1, \ldots, x_n]$ and that, for any substitution of values $x_1 = s_1, x_2 = s_2, \ldots, x_n = s_n$ of the x_i in I such that each $g_i(s_1, \ldots, s_n) \neq 0$, we have $f(s_1, \ldots, s_n) = 0$. Then $f = 0 \in I[x_1, \ldots, x_n]$.*

The proof in the case of one indeterminate follows from Section 21 of *Modern Algebra* by van der Waerden. One then proves the general case by induction on n.

Suppose now that $x_1, \ldots, x_m$ are indeterminates over K. Then $K[x_1, \ldots, x_m]$ is a unique factorization domain, and hence any two polynomials $p_1, p_2 \in K[x_1, \ldots, x_m]$ have a greatest common divisor. Any two greatest common divisors differ by a unit, that is, an element $\neq 0$ of K. Furthermore, in a unique factorization domain, if p_1 and p_2 are relatively prime, that is, 1 is a g.c.d. of p_1 and p_2, and p_1 divides qp_2, then p_1 divides q.

A choice of basis $e_1, \ldots, e_m$ of E determines coordinate functions $\xi_1, \ldots, \xi_m : E \to K$ defined for each $i = 1, \ldots, m$ by $\xi_i(\Sigma\lambda_j e_j) = \lambda_i$. The ring of polynomial functions on E is defined to be the subring $S(E)$ of $\mathscr{F}(E, K)$ generated by the coordinate functions $\xi_1, \ldots, \xi_m$. By the principle of irrelevancy $S(E)$ is isomorphic to $K[x_1, \ldots, x_m]$ by the degree preserving isomorphism which sends $\xi_i \to x_i$, for $i = 1, \ldots, m$, and sends the constant polynomial λ to $\lambda \in K \subset K[x_1, \ldots, x_m]$. Hence the ring of polynomial functions on E is a unique factorization domain.

Lemma. *If $p_1(y), \ldots, p_r(y)$ are polynomial functions on E with g.c.d. $= 1$, then for any $\sigma \in \Gamma$, $p_1(\sigma \cdot y), \ldots, p_r(\sigma \cdot y)$ are polynomial functions on E with g.c.d. $= 1$.*

Proof. For the map $y \to \sigma \cdot y$, being a linear automorphism of E, determines a *degree preserving* automorphism of the ring of polynomial functions on E by $p(y) \to p(\sigma \cdot y)$. Thus if $q(y)$ is a non-constant divisor of $p_1(\sigma \cdot y), \ldots, p_r(\sigma \cdot y)$, then $q(\sigma^{-1} \cdot y)$ is a non-constant divisor of $p_r(y), \ldots, p_r(y)$.

This ends our algebraic digression.

Proposition 1. *Every non-zero rational concomitant $u : E \to F$ can be expressed as $u = (p/q)\, v$, where $v : E \to F$ is a polynomial concomitant and p, q are polynomial relative invariants.*

Proof. Express $u(x) = \Sigma_k u_k(x) f_k$ where the u_k are rational functions on E. By the above remarks, we may write each u_k as $(p/q)\, v_k$ where p, q are relatively prime polynomials and the v_k are polynomials such that the g.c.d. of $v_1, \ldots, v_n$ is 1. For $\sigma \in \Gamma$, the concomitance of u implies that for each $x \in E$ such that $q(x) \neq 0$ and $q(\sigma \cdot x) \neq 0$, we have

$$p(x)\, q(\sigma \cdot x)\, \Sigma_j a_{ij}(\sigma)\, v_q(x) = p(\sigma \cdot x)\, q(x)\, v_i(\sigma \cdot x) \tag{1}$$

for $i = 1, \ldots, n$, where $(a_{ij}(\sigma))$ is the matrix associated to $\mu_\sigma : F \to F$ by the basis $f_1, \ldots, f_n$. Since the field K is infinite, the principle of irrelevancy (suitably reformulated for the ring of polynomials on E) implies that (1) holds for every $x \in E$.

Since p divides the left hand side of the equation for each i, and since the $v_i(\sigma \cdot x)$ have no common factor, $p(x)$ divides $p(\sigma \cdot x)$. Since $p(x)$ and $p(\sigma \cdot x)$ have the same degree, $p(\sigma \cdot x) = d(\sigma)\, p(x)$, where $d : \Gamma \to K^*$ is a character. Likewise, $q(\sigma \cdot x)$ divides $q(x)$, and hence $q(\sigma \cdot x) = d'(\sigma)\, q(x)$ where $d' : \Gamma \to K^*$ is another character. Consequently,

$v_i(\sigma \cdot x) = \chi(\sigma) \Sigma_j a_{ij}(\sigma) v_j(x)$ with $\chi(\sigma) = d'(\sigma)/d(\sigma)$. This expresses precisely the fact that $v = \Sigma v_i f_i$ is a polynomial concomitant for the linear action $(\sigma, y) \to \chi(\sigma) \sigma \cdot y$ of Γ on F.

We have thus reduced the determination of rational concomitants to the determination of *polynomial* concomitants. We have further

Proposition 2. *Let* $u : E \to F$ *be a polynomial concomitant and write* $u = u_0 + u_1 + \cdots + u_k$, *where each* u_j *is homogeneous of degree* j. *Then each* u_j *is a concomitant.*

Proof. Since $u(\sigma \cdot x) = \sigma \cdot u(x)$ for all $x \in E$, it follows that

$$\begin{aligned} u_0 &+ \lambda u_1(\sigma \cdot x) + \cdots + \lambda^k u_k(\sigma \cdot x) \\ &= \sigma \cdot u_0 + \lambda\sigma \cdot u_1(x) + \cdots + \lambda^k \sigma \cdot u_k(x) \end{aligned}$$

for every $\lambda \in K$. As K is infinite, this is only possible if $\sigma \cdot u_i(x) = u_i(\sigma \cdot x)$ for each i and $x \in E$.

The polarization process

Let $\xi_1, \ldots, \xi_n$ be n indeterminates over the field K.

Definition. A homogeneous polynomial in $K[\xi_1, \ldots, \xi_n]$ of total degree r is called an *n-ary form of order r*.

Let f be an n-ary form of order r, say $f(x) = f(\xi_1, \ldots, \xi_n)$. Let $z = (\zeta_1, \ldots, \zeta_n) \in K^n$. By Taylor's formula,

$$f(x + \lambda z) = f(x) + \lambda D_{zx} f(x) + \cdots$$

where

$$D_{zx} f(x) = \zeta_1 \frac{\partial f}{\partial \xi_1}(x) + \cdots + \zeta_n \frac{\partial f}{\partial \xi_n}(x).$$

Since f is homogeneous of degree r, Euler's theorem implies

$$D_{xx} f(x) = r \cdot f(x),$$

and hence $f(x)$ can be restituted from $D_{zx} f(x)$. $D_{zx} f(x)$ is called the first partial polarization of f. Since f is of degree r, the function

$$Pf(z^{(1)}, \ldots, z^{(r)}) = D_{z^{(r)}x} D_{z^{(r-1)}x} \cdots D_{z^{(1)}x} f(x)$$

is a function of $z^{(1)}, \ldots, z^{(r)}$ only and is in fact multilinear and symmetric in $z^{(1)}, \ldots, z^{(r)}$. It is clear how to define Pf for a vector valued polynomial f.

Definition. Let f be a vector valued n-ary form of order r. Then Pf is called the *total polarization* of f.

The restitution process.

By successive application of Euler's theorem, one obtains $Pf(x, \ldots, x) = r! f(x)$. Since the field K has characteristic 0, $f(x)$ may be restituted from $Pf(x, \ldots, x)$; namely, $f(x) = (r!)^{-1} Pf(x, \ldots, x)$.

Proposition 3. *Let $u : E \to F$ be a homogeneous polynomial concomitant of degree r. Then if $(x_1, \ldots, x_r) \to v(x_1, \ldots, x_r) : E^r \to F$ is the polarized mapping, v is a multilinear simultaneous concomitant of $x_1, \ldots, x_r$.*

Proof. Using the standard multi-index notations, we have, for $\lambda_1, \ldots, \lambda_r \in K$, the representation

$$u(\lambda_1 x_1 + \cdots + \lambda_r x_r) = \sum_{\substack{\alpha=(\alpha_1,\ldots,\alpha_n) \\ |\alpha|=r}} c_\alpha \lambda^\alpha u_\alpha(x_1, \ldots, x_r).$$

Applying $\sigma \in \Gamma$ and using the concomitance of u,

$$u(\lambda_1 \sigma \cdot x_1 + \cdots + \lambda_r \sigma \cdot x_r) = \sigma \cdot u(\lambda_1 x_1 + \cdots + \lambda_r x_r),$$

we obtain

$$\Sigma_\alpha c_\alpha \lambda^\alpha u_\alpha(\sigma \cdot x_1, \ldots, \sigma \cdot x_r) = \Sigma_\alpha c_\alpha \lambda^\alpha \sigma \cdot u_\alpha(x_1, \ldots, x_r).$$

Since the field K is infinite, $\sigma \cdot u_\alpha(x_1, \ldots, x_r) = u_\alpha(\sigma \cdot x_1, \ldots, \sigma \cdot x_r)$ for each α, so each u_α is a simultaneous concomitant. Since v is one of the u_α, in fact $v = u_{(1,\ldots,1)}$, v is indeed a multilinear simultaneous concomitant.

The final reduction of the problem from multilinear concomitance to linear concomitance follows from multilinear algebra in the way one obtains a linear map from a multilinear map.

Let $E_1, \ldots, E_s$, F be finite dimensional K-vector spaces, and let $\mathscr{L}(E_1, \ldots, E_s; F)$ denote the space of all multilinear maps

$$u : E_1 \times \cdots \times E_s \to F.$$

If each $E_i = E$, we denote this by $\mathscr{L}_s(E; F)$. Let $E_1 \otimes \cdots \otimes E_s$ be the tensor product (over K) of $E_1, \ldots, E_s$, which becomes, in our notation, $E^{\otimes s}$, the s-th tensor power of E, if each $E_i = E$. The natural map $\varphi : E_1 \times \cdots \times E_s \to E_1 \otimes \cdots \otimes E_s$ induces an isomorphism

$$\operatorname{Hom}(E_1 \otimes \cdots \otimes E_s, F) \to \mathscr{L}(E_1, \ldots, E_s; F) \qquad \text{by} \quad w \to w\varphi.$$

($\operatorname{Hom}(E, F)$ represents the space of all K-linear maps from E to F.)

Suppose Γ acts on $E_1, \ldots, E_s$. Then Γ acts on $E_1 \otimes \cdots \otimes E_s$ by defining $\sigma \cdot (x_1 \otimes \cdots \otimes x_s) = \sigma \cdot x_1 \otimes \cdots \otimes \sigma \cdot x_s$ on the generators $x_1 \otimes \cdots \otimes x_s$ and requiring the action of Γ to be linear. This action is well defined, by the universal property of the tensor product, since $(x_1, \ldots, x_s) \to \sigma \cdot x_1 \otimes \cdots \otimes \sigma \cdot x_s$ is multilinear. The corresponding representation is called a *tensor representation* of Γ.

Proposition 4. *If $v \in \mathscr{L}_r(E; F)$ is a simultaneous multilinear concomitant and $v = w\varphi$ where $w \in \operatorname{Hom}(E^{\otimes r}, F)$, then w is a linear concomitant and conversely.*

The proof is immediate from the definitions.

Chapter 2. Rational Representations of the General Linear Group

1. Representations of Linear Groups

Let Γ be a group acting linearly on a finite dimensional K-vector space E. Equivalently, suppose there is given a linear representation of Γ in $\mathbf{GL}(E)$. Then E is called a *Γ-module*. If F is another finite dimensional Γ-module, then a linear concomitant $u : E \to F$ for Γ is, by definition, a Γ-module homomorphism.

Definition. A finite dimensional Γ-module E is called *simple*, and the corresponding representation is called *irreducible*, if there exists no non-trivial proper Γ-invariant subspace of E. If E is a direct sum of simple Γ-modules, the representation of Γ in $\mathbf{GL}(E)$ is said to be *completely reducible*.

Definition. If E is a simple Γ-module, then any concomitant u for Γ defined on E is called an *irreducible* concomitant.

Suppose now that Γ is a subgroup of $\mathbf{GL}(n; K)$, the group of non-singular $n \times n$ matrices over K. *We shall always assume that whenever Γ acts on a finite dimensional K-vector space E, it acts rationally*. In other words, the representation $\underline{F} : \Gamma \to \mathbf{GL}(E)$ is such that if the elements of $\mathbf{GL}(E)$ are written as matrices (with respect to some basis of E), then the components of $\underline{F}(\underline{X})$ are rational functions of the components x_{ij} of $\underline{X}$. Such a representation is called a *rational linear representation* of Γ. Γ will be called *reductive* if and only if every finite dimensional *rational* representation of Γ is completely reducible.

Notation. The algebra of $n \times n$ matrices over K will be denoted by $\underline{M}_n(K)$. Matrices will always be denoted by underlined capital letters $\underline{X}$, $\underline{Y}$, etc.

We will now consider the important special case $\Gamma = \mathbf{GL}(n; \Omega)$, where Ω is an algebraically closed field of characteristic zero. Suppose $\underline{X} \to \underline{F}(\underline{X})$ is a rational linear representation of $\mathbf{GL}(n; \Omega)$ in $\mathbf{GL}(m; \Omega)$. Then the relation $\underline{F}(\underline{X}\underline{Y}) = \underline{F}(\underline{X})\underline{F}(\underline{Y})$ holds for all $\underline{X}, \underline{Y} \in \underline{M}_n(\Omega)$, provided $\det(\underline{X}) \neq 0$ and $\det(\underline{Y}) \neq 0$. Since $\underline{F}$ is rational, we may express $\underline{F}(\underline{X}) = (p(\underline{X})/q(\underline{X}))\, \underline{G}(\underline{X})$, where p and q are relatively prime polynomials and $\underline{G}$ is a matrix with polynomial entries whose greatest common divisor is 1. The equation $\underline{F}(\underline{X}\underline{Y}) = \underline{F}(\underline{X})\underline{F}(\underline{Y})$ implies

$$q(\underline{X}\underline{Y})\,p(\underline{X})\,p(\underline{Y})\,\underline{G}(\underline{X})\,\underline{G}(\underline{Y}) = q(\underline{X})\,q(\underline{Y})\,p(\underline{X}\underline{Y})\,\underline{G}(\underline{X}\underline{Y}), \tag{1}$$

provided det($\underline{X}$) and det($\underline{Y}$)are both non-zero. Hence (1) holds provided a finite number of polynomial inequalities hold. By the *principle of irrelevancy of algebraic inequalities*, (1) must therefore be valid for all $\underline{X}$, $\underline{Y}$ in $\underline{M}_n(\Omega)$.

Proposition 1. *Suppose $\underline{X} \to \underline{F}(\underline{X})$ is a rational linear representation of* $\mathbf{GL}(n; \Omega)$. *Then $\underline{F} = (p/q)\,\underline{G}$, where p and q are polynomial characters, and $\underline{G}$ is a polynomial linear representation of* $\mathbf{GL}(n; \Omega)$.

Proof. As above, express $\underline{F} = (p/q)\,\underline{G}$ with p and q normalized so that $p(\underline{I}_n) = 1$ and $q(\underline{I}_n) = 1$. Then

$$p(\underline{X}\underline{Y})\,q(\underline{X})\,q(\underline{Y})\,\underline{G}(\underline{X}\underline{Y}) = p(\underline{X})\,p(\underline{Y})\,q(\underline{X}\underline{Y})\,\underline{G}(\underline{X})\,\underline{G}(\underline{Y}) \tag{1}$$

for all $\underline{X}, \underline{Y} \in \underline{M}_n(\Omega)$. By the usual reasoning, (1) still obtains if we suppose $\underline{X}$ and $\underline{Y}$ are matrices of n^2 indeterminates. Extend Ω to $\Omega(\underline{Y})$. Since $\underline{G}(\underline{I}_n) = \underline{I}_m$, det $\underline{G}(\underline{Y}) \neq 0$, and thus we may multiply both sides of (1) by $\underline{G}(\underline{Y})^{-1}$ to obtain

$$p(\underline{X}\underline{Y})\,q(\underline{X})\,q(\underline{Y})\,\underline{G}(\underline{X}\underline{Y})\,\underline{G}(\underline{Y})^{-1} = p(\underline{X})\,p(\underline{Y})\,q(\underline{X}\underline{Y})\,\underline{G}(\underline{X}). \tag{2}$$

We see immediately that $q(\underline{X})$ divides $q(\underline{X}\underline{Y})$ in $\Omega(\underline{Y})[\underline{X}]$, and hence, by Gauss' Lemma (see "Modern Algebra" by van der Waerden), $q(\underline{X})$ divides $q(\underline{X}\underline{Y})$ in $\Omega[\underline{Y}][\underline{X}] = \Omega[\underline{X}, \underline{Y}]$. Similarly, $q(\underline{Y})$ divides $q(\underline{X}\underline{Y})$. Since $q(\underline{X})$ and $q(\underline{Y})$ are relatively prime, we conclude that $q(\underline{X}\underline{Y}) = f(\underline{X}, \underline{Y})\,q(\underline{X})\,q(\underline{Y})$ for some $f(\underline{X}, \underline{Y}) \in \Omega[\underline{X}, \underline{Y}]$.

f is, in fact, a polynomial of degree zero. For suppose that q has total degree d. Then $q(\underline{X})\,q(\underline{Y})$ has total degree $2d$. But $\underline{X}\underline{Y}$ is a family of n^2 polynomials of degree 2, and this implies that $q(\underline{X}\underline{Y})$ must have total degree $\leqslant 2d$. As this is only possible if f has degree 0, the assertion is established. Since $q(\underline{I}_n) = 1$, we conclude that $q(\underline{X}\underline{Y}) = q(\underline{X})\,q(\underline{Y})$.

By considering the (contragredient) representation $\underline{X} \to {}^t\underline{F}^{-1}(\underline{X})$ (which interchanges the roles of p and q) we see that $p(\underline{X}\underline{Y}) = p(\underline{X})\,p(\underline{Y})$. From this it follows that $\underline{G}(\underline{X}\underline{Y}) = \underline{G}(\underline{X})\,\underline{G}(\underline{Y})$, so p, q, and $\underline{G}$ satisfy the requirements of the proposition.

We have shown that every rational linear representation of $\mathbf{GL}(n; \Omega)$ is, up to a rational character, a polynomial representation. We will now show that every polynomial representation of $\mathbf{GL}(n; \Omega)$ in $\mathbf{GL}(E)$ is, after composing with an inner automorphism of $\mathbf{GL}(E)$, a direct sum of *homogeneous* polynomial representations. To show this, we will appeal to the Jordan canonical form of a matrix, and thus the assumption that Ω is algebraically closed is needed.

If $\underline{U}$ is any $n \times n$ matrix over an algebraically closed field Ω, there exists a nonsingular matrix $\underline{S}$ over Ω such that $\underline{S}\underline{U}\underline{S}^{-1}$ is the direct sum of matrices of the form

$$\begin{pmatrix} \alpha & 1 & 0 & \cdots & 0 \\ & \alpha & 1 & \cdots & 0 \\ & & \ddots & \ddots & \\ & & & \alpha & 1 \\ & & & & \ddots \\ 0 & & & & \alpha \end{pmatrix}$$

with α's on the diagonal, 1's directly above the diagonal, and 0's elsewhere. Each α appearing on the diagonal of $\underline{S}\underline{U}\underline{S}^{-1}$ is a root of the minimal polynomial of $\underline{U}$. A matrix in this form is said to be in *Jordan canonical form.*

Now suppose $\underline{F} : \mathbf{GL}(n; \Omega) \to \mathbf{GL}(m; \Omega)$ is a polynomial representation, and consider the restriction of $\underline{F}$ to the center of $\mathbf{GL}(n; \Omega)$. Thus, for $\lambda \in \Omega$, define $\underline{G}(\lambda) = \underline{F}(\lambda \underline{I}_n)$. If $\mu \in \Omega$ also, we have $\underline{G}(\lambda + \lambda\mu) = \underline{G}(\lambda)\, \underline{G}(1 + \mu)$, so, by Taylor's theorem,

$$\underline{G}(\lambda)(\underline{I}_m + \mu \underline{G}'(1) + \cdots) = \underline{G}(\lambda) + \lambda\mu \underline{G}'(\lambda) + \cdots .$$

Since Ω is infinite, this can happen only if $\underline{G}(\lambda)\, \underline{G}'(1) = \lambda \underline{G}'(\lambda)$.

Let $\underline{G}'(1) = \underline{P}$. We may assume $\underline{P}$ is in Jordan canonical form, because if $\underline{S} \in \mathbf{GL}(m; \Omega)$ is such that $\underline{S}\underline{P}\underline{S}^{-1}$ is in Jordan canonical form, then $\underline{X} \to \underline{S}\underline{F}(\underline{X})\, \underline{S}^{-1}$ is a polynomial representation *equivalent* to $\underline{F}$ with $\underline{G}'(1)$ in Jordan canonical form. We assert that $\underline{P}$ is a diagonal matrix. If not, after another similarity transformation $\underline{P}$ is necessarily of the form

$$\begin{pmatrix} \alpha & 1 & \cdots \\ 0 & \alpha & \\ \vdots & & \ddots \end{pmatrix}.$$

Assume $\underline{G}(\lambda) = (f_{ij}(\lambda))$. Then from $\underline{G}(\lambda)\,\underline{P} = \lambda\underline{G}'(\lambda)$ we obtain

$$\lambda f'_{11}(\lambda) = \alpha f_{11}(\lambda) \tag{3}$$

$$f_{11}(\lambda) + \alpha f_{12}(\lambda) = \lambda f'_{12}(\lambda). \tag{4}$$

From (3) we see that $\alpha = \deg f_{11}$, say $\alpha = j$. If f_{12} has degree k, we obtain from (4) that

$$a\lambda^j = -b(j-k)\,\lambda^k$$

where $f_{11}(\lambda) = a\lambda^j + \cdots$ and $f_{12}(\lambda) = b\lambda^k + \cdots$ with $ab \neq 0$. Since Ω has characteristic 0, this is impossible so $\underline{P}$ must be diagonal. Furthermore, this argument repeated implies the diagonal entries of $\underline{P}$ are positive integers. Hence we may write (after another similarity)

$$\underline{P} = \begin{pmatrix} m_1\underline{I}_{m_1} & & 0 \\ & \ddots & \\ 0 & & m_p\underline{I}_{m_p} \end{pmatrix}$$

with $m_1 > m_2 > \cdots > m_p > 0$ all integers. It follows that

$$\underline{G}(\lambda) = \begin{pmatrix} \lambda^{m_1}\underline{I}_{m_1} & & 0 \\ & \ddots & \\ 0 & & \lambda^{m_p}\underline{I}_{m_p} \end{pmatrix}$$

since $\underline{G}(\lambda)\,\underline{P} = \lambda\underline{G}'(\lambda)$.

Suppose now that

$$\underline{F}(\underline{X}) = \begin{pmatrix} \underline{F}_{11}(\underline{X}) & \cdots & \underline{F}_{1m}(\underline{X}) \\ \vdots & & \vdots \\ \underline{F}_{m1}(\underline{X}) & \cdots & \underline{F}_{mm}(\underline{X}) \end{pmatrix}$$

where $\underline{F}_{ij}(\underline{X})$ is an $m_i \times m_j$ matrix. Since $\underline{F}(\underline{X})\,\underline{G}(\lambda) = \underline{G}(\lambda)\,\underline{F}(\underline{X}) = \underline{F}(\lambda\underline{X})$, we have, for $i \neq j$, $\lambda^{m_i}\underline{F}_{ij}(\underline{X}) = \lambda^{m_j}\underline{F}_{ij}(\underline{X})$, and since $m_i \neq m_j$, $\underline{F}_{ij}(\underline{X}) = 0$. Furthermore, $\underline{F}_{ii}(\lambda\underline{X}) = \lambda^{m_i}\underline{F}_{ii}(\underline{X})$ so that $\underline{F}_{ii}(\underline{X})$ is a *homogeneous* polynomial representation of $\mathbf{GL}(n; \Omega)$ in $\mathbf{GL}(m_i; \Omega)$. Therefore,

Theorem. *A polynomial linear representation of* $\mathbf{GL}(n; \Omega)$ *is always equivalent to a direct sum of homogeneous polynomial linear representations.*

2. Representations of the Full Linear Group

Let $\underline{F}$ be a homogeneous polynomial representation of the full linear group $\Gamma = \mathbf{GL}(n; \Omega)$ over an algebraically closed field of characteristic zero. We shall prove that $\underline{F}$ can be lifted uniquely from Γ to a homomorphism of a certain semi-simple algebra $A_n{}^f$ (f is the degree of the representation) in which Γ is imbedded. It will follow that Γ is reductive. The key is to study the relationship between the homogeneous representations of Γ and certain representations of the symmetric groups. Throughout this section we will depend on results from the Appendix.

Let $e_1 ,..., e_n$ be a basis of E ($\cong \Omega^n$), and, for each multi-index $\alpha = (\alpha_1 ,..., \alpha_f) \in I^f$ $(I = \{1, 2,..., n\})$, set $e_\alpha = e_{\alpha_1} \otimes e_{\alpha_2} \otimes \cdots \otimes e_{\alpha_f}$. Then the vectors e_α as α ranges over I^f form a basis of $E^{\otimes f}$. Let $\mathcal{G}_f$ denote the symmetric group on f letters. If, for $\pi \in \mathcal{G}_f$ and $\alpha \in I^f$, we set $\pi \cdot \alpha = (\alpha_{\pi^{-1}(1)} , \alpha_{\pi^{-1}(2)} ,..., \alpha_{\pi^{-1}(f)})$, then the relation $\pi \cdot e_\alpha = e_{\pi.\alpha}$ determines a well defined linear action of $\mathcal{G}_f$ on $E^{\otimes f}$. It is natural therefore to single out the linear concomitants of this action. Let End(E) denote Hom(E, E).

Definition. $\underline{U} \in \mathrm{End}(E^{\otimes f})$ is called *bisymmetric* if $\pi\underline{U} = \underline{U}\pi$ for all $\pi \in \mathcal{G}_f$.

In other words, $\underline{U}$ is bisymmetric if and only if $\underline{U}$ is a concomitant of $\mathcal{G}_f$ or if and only if $\underline{U}$ is an element of the commutant of the image of $\Omega[\mathcal{G}_f]$ in $\mathrm{End}(E^{\otimes f})$, $\Omega[\mathcal{G}_f]$ being the group ring of $\mathcal{G}_f$ over Ω. Now Maschke's theorem says that $\Omega[\mathcal{G}_f]$, and hence its image in $\mathrm{End}(E^{\otimes f})$, is semi-simple. Therefore so is the algebra $A_n{}^f$ of bisymmetric endomorphisms of $E^{\otimes f}$ by Schur's commutation theorem. This means that every finite dimensional $A_n{}^f$-module decomposes as a direct sum of simple $A_n{}^f$-submodules; i.e., subspaces invariant under $A_n{}^f$ each one having no proper invariant subspace. We shall presently apply this result to representation theory.

Proposition. *Suppose* $\underline{U} \in \mathrm{End}(E^{\otimes f})$ *has matrix* $(u_{\alpha\beta})$ *where* $\alpha, \beta \in I^f$ (*indices are ordered lexicographically*). *Then* $\underline{U}$ *is bisymmetric if and only if* $u_{\alpha\beta} = u_{\pi\cdot\alpha,\pi\cdot\beta}$ *for each* $\pi \in \mathcal{G}_f$.

Proof. From $\underline{U}(\pi \cdot e_\alpha) = \pi \cdot \underline{U}(e_\alpha)$ and

$$\underline{U}(e_{\pi\cdot\alpha}) = \sum_\beta u_{\pi\cdot\alpha,\beta} e_\beta$$

$$= \sum_\beta u_{\pi\cdot\alpha,\pi\cdot\beta} e_{\pi\cdot\beta}$$

we get

$$\begin{aligned}\underline{U}(e_\alpha) &= \pi^{-1} \cdot \underline{U}(e_{\pi\cdot\alpha}) \\ &= \sum_\beta u_{\pi\cdot\alpha,\pi\cdot\beta} e_\beta .\end{aligned}$$

Therefore $u_{\pi\cdot\alpha,\pi\cdot\beta} = u_{\alpha\beta}$ for all π. The converse is similar.

Due to the isomorphism $\mathrm{End}(E^{\otimes f}) \cong \mathrm{End}(E)^{\otimes f}$ (gotten by mapping $\underline{U}_1 \otimes \cdots \otimes \underline{U}_f$ to the endomorphism $\underline{U}$ such that $\underline{U}(y_1 \otimes \cdots \otimes y_f) = \underline{U}_1(y_1) \otimes \cdots \otimes \underline{U}_f(y_f)$) we may consider the tensor representations $\Gamma \to GL(E^{\otimes f})$ defined by $\underline{X} \to \underline{X}^{\otimes f}$ for $\underline{X} \in \Gamma$. Suppose that $\underline{X} \in \mathrm{End}(E)$ has matrix (x_{ij}). Then one sees without difficulty that $\underline{X}^{\otimes f}$ has matrix $(X_{\alpha\beta})$, where $X_{\alpha\beta} = x_{\alpha_1\beta_1} \cdots x_{\alpha_f\beta_f}$ for each pair of multi-indices $\alpha = (\alpha_1, \ldots, \alpha_f)$ and $\beta = (\beta_1, \ldots, \beta_f)$ in I^f. Since $\underline{X}^{\otimes f}$ is clearly bisymmetric, $X_{\alpha\beta} = X_{\pi\cdot\alpha,\pi\cdot\beta}$ for each $\pi \in \mathscr{G}_f$.

Lemma. *Let x_{ij} $(1 \leqslant i, j \leqslant n)$ be algebraically independent over Ω. Then, for any $\alpha, \beta, \epsilon, \mu \in I^f$, $X_{\alpha\beta} = X_{\epsilon\mu}$ if and only if there exists a permutation $\pi \in \mathscr{G}_f$ such that $\pi \cdot \alpha = \epsilon$ and $\pi \cdot \beta = \mu$.*

Proof. We proceed by induction on f. The case $f = 1$ is trivial. Continuing on, write $X_{\alpha\beta} = x_{\alpha_1\beta_1} \cdots x_{\alpha_f\beta_f}$ and $X_{\epsilon\mu} = x_{\epsilon_1\mu_1} \cdots x_{\epsilon_f\mu_f}$. There must exist a k such that $x_{\epsilon_k\mu_k} = x_{\alpha_f\beta_f}$. Let π_0 be the transposition interchanging k and f. Then $X_{\pi_0\cdot\epsilon,\pi_0\cdot\mu} = (x_{\epsilon_1\mu_1} \cdots x_{\epsilon_f\mu_f} \cdots x_{\epsilon_{f-1}\mu_{f-1}})x_{\epsilon_k\mu_k}$. Now consider new multi-indices $\epsilon', \mu' \in I^{f-1}$ where $\epsilon_i' = \epsilon_i$ if $i \neq k$ and $i < f$ and $\epsilon_k' = \epsilon_f$ with μ' being defined in the same manner. Let $\alpha' = (\alpha_1, \ldots, \alpha_{f-1})$ and $\beta' = (\beta_1, \ldots, \beta_{f-1})$. By the induction hypothesis, there must exist a $\pi' \in \mathscr{G}_{f-1}$ such that $\pi' \cdot \alpha' = \epsilon'$ and $\pi' \cdot \beta' = \mu'$. If we extend π' to $\mathscr{G}_f$ by $\pi'(f) = f$, it follows that $\pi' \cdot \alpha = \pi_0 \cdot \epsilon$ and $\pi' \cdot \beta = \pi_0 \cdot \mu$.

For a homogeneous (polynomial) representation $\underline{F} : \Gamma \to \mathbf{GL}(N; \Omega)$ of degree f we may write each entry $F_{hk}(\underline{X}) = \sum_{\alpha,\beta} a_{hk\alpha\beta} X_{\alpha\beta}$ (summation over all $\alpha, \beta \in I^f$). By the last lemma, if we require that $a_{hk\alpha\beta} = a_{hk,\pi\cdot\alpha,\pi\cdot\beta}$ for all $\pi \in \mathscr{G}_f$, then this representation is unique. Letting $\underline{M}_N(\Omega)$ denote the algebra of $N \times N$ matrices over Ω, we define a linear mapping $\underline{G} : A_n^f \to \underline{M}_N(\Omega) = \mathrm{End}(\Omega^N)$ by $G_{hk}(\underline{U}) = \sum_{\alpha,\beta} a_{hk\alpha\beta} u_{\alpha\beta}$.

$\underline{G}$ is in fact a homomorphism. For, first of all, $\underline{F}$ being polynomial, the relation $\underline{F}(\underline{X}\underline{Y}) = \underline{F}(\underline{X})\underline{F}(\underline{Y})$ for all $\underline{X}, \underline{Y} \in \Gamma$ must hold, so, by principle of irrelevancy, for all $\underline{X}, \underline{Y} \in \mathrm{End}(E)$, the irrelevant inequalities being $\det(\underline{X}) \neq 0$ and $\det(\underline{Y}) \neq 0$. Now $\underline{F}(\underline{X}) = \underline{G}(\underline{X}^{\otimes f})$, and since $(\underline{X}\underline{Y})^{\otimes f} = \underline{X}^{\otimes f}\underline{Y}^{\otimes f}$ ($\underline{X} \to \underline{X}^{\otimes f}$ is a representation) we certainly have

$\underline{G}(\underline{X}^{\otimes f}\underline{Y}^{\otimes f}) = \underline{F}(\underline{X}\underline{Y}) = \underline{F}(\underline{X})\underline{F}(\underline{Y}) = \underline{G}(\underline{X}^{\otimes f})\,\underline{G}(\underline{Y}^{\otimes f})$. The proof of the assertion will thus be complete once the following lemma is established.

Lemma. *A_n^f is the subalgebra of* $\mathrm{End}(E^{\otimes f})$ *generated by all $\underline{X}^{\otimes f}$ for* $X \in \mathrm{End}(E)$.

Proof. Suppose there exists a relation $\sum_{\alpha,\beta} b_{\alpha\beta}X_{\alpha\beta} = 0$ valid for all $\underline{X} \in \mathrm{End}(E)$ where $b_{\pi\cdot\alpha,\pi\cdot\beta} = b_{\alpha\beta}$ for all $\pi \in \mathscr{G}_f$. Then certainly each $b_{\alpha\beta} = 0$ since Ω has characteristic zero. But if the $\underline{X}^{\otimes f}$ generate the algebra B, this relation says that the only vector $(b_{\alpha\beta}) \in A_n^f$ perpendicular to B is $(b_{\alpha\beta}) = 0$, and that is possible only if $B = A_n^f$.

Theorem. *Any homogeneous polynomial representation*

$$\underline{F} : \mathbf{GL}(n; \Omega) \to \mathbf{GL}(N; \Omega)$$

of degree f factors uniquely as $\underline{X} \to \underline{X}^{\otimes f} \to \underline{G}(\underline{X}^{\otimes f})$ where $\underline{G}$ is a homomorphism of the algebra A_n^f into $\underline{M}_N(\Omega)$.

Proof. The only assertion yet unproven is the uniqueness of $\underline{G}$ which follows from the previous lemma.

Note that this theorem is valid for any field of characteristic zero. The next theorem is the main theorem of this section.

Theorem. *The full linear group* $\mathbf{GL}(n; \Omega)$ *over an algebraically closed field of characteristic zero is reductive; that is, every rational linear representation of* $\mathbf{GL}(n; \Omega)$ *is completely reducible.*

Proof. From the previous section we may replace rational by polynomial and polynomial by homogeneous polynomial. But for homogeneous polynomial representations the proof follows from the last theorem and the fact that each A_n^f is semi-simple.

3. Young's Diagrams

We now turn to the problem of determining all characters or 1-dimensional representations of $\mathbf{GL}(n; \Omega)$. Every 1-dimensional polynomial representation of $\mathbf{GL}(n; \Omega)$ is by § 1 homogeneous of degree f, and thus it induces a 1-dimensional representation of A_n^f which must give, in particular, a simple A_n^f-module. By Part IV of the Appendix, *every* simple A_n^f-module is isomorphic to one of the form $b \cdot E^{\otimes f}$, where b generates a minimal left ideal of the image of $\Omega[\mathscr{G}_f]$ in $\mathrm{End}(E^{\otimes f})$. Hence we are led to consider two problems:

(i) Classify all minimal left ideals of $\Omega[\mathscr{G}_f]$ (with respect to their generators); and

(ii) using (i), determine all A_n^f-submodules of $E^{\otimes f}$ of dimension 1 (over Ω).

In this section we shall consider the first problem, whose solution is due to Frobenius and A. Young (1901–1903). Our discussion is based on the presentations of J. von Neumann and van der Waerden.

Definition. A *Young's frame* α is a sequence $\alpha = (\alpha_1, ..., \alpha_r)$ of integers such that $\alpha_1 \geqslant \alpha_2 \geqslant ... \geqslant \alpha_r \geqslant 1$. The *number of cases* of a frame α is defined to be the integer $f = \sum_{i=1}^{r} \alpha_i$. *A Young's diagram* Σ_α corresponding to a frame α is a double sequence (m_{ij_i}) $(1 \leqslant i \leqslant r, 1 \leqslant j_i \leqslant \alpha_i)$ of distinct integers between 1 and f, the number of cases.

We shall only consider frames of a fixed number f of cases. One notes that a diagram Σ_α may be arranged in an array

$$\begin{matrix} m_{11} & m_{12} & \cdots & m_{1\alpha_1} \\ m_{21} & m_{22} & \cdots & m_{2\alpha_2} \\ \vdots & & & \\ m_{r1} & \cdots & m_{r\alpha_r} & \end{matrix}$$

$\mathscr{G}_f$ acts on a diagram Σ_α by defining $s \cdot \Sigma_\alpha$ $(s \in \mathscr{G}_f)$ to be the diagram (m'_{ij_i}) (corresponding to α) where $m'_{ij_i} = s \cdot m_{ij_i}$.

There are two important subgroups of $\mathscr{G}_f$ determined by Σ_α. The first, $\mathscr{R}(\Sigma_\alpha)$, is the subgroup of $\mathscr{G}_f$ that leaves invariant each row of Σ_α, and the second, $\mathscr{C}(\Sigma_\alpha)$, is the subgroup of $\mathscr{G}_f$ which leaves invariant each column of Σ_α. We note the following facts:

(i) $\mathscr{R}(\Sigma_\alpha) \cap \mathscr{C}(\Sigma_\alpha) = \{e\}$;

(ii) if $s \in \mathscr{G}_f$, then $\mathscr{R}(s \cdot \Sigma_\alpha) = s\mathscr{R}(\Sigma_\alpha)s^{-1}$, and $\mathscr{C}(s \cdot \Sigma_\alpha) = s\mathscr{C}(\Sigma_\alpha)s^{-1}$.

Order the frames α lexicographically. That is, say $\alpha' > \alpha$ if $\alpha_i' > \alpha_i$ for the least i such that $\alpha_i' \neq \alpha_i$. The following combinatorial lemma, due to J. von Neumann, is the fundamental lemma of our discussion.

Lemma. *Let* Σ_α, Σ_β' *be two Young's diagrams such that* $\alpha \geqslant \beta$. *Then either*:

(i) *there exist distinct integers in* $\{1, ..., f\}$ *in the same row of* Σ_α *and the same column of* Σ_β', *or*

(ii) $\alpha = \beta$ *and* $\Sigma_\alpha' = pq \cdot \Sigma_\alpha$ *for some* $p \in \mathscr{R}(\Sigma_\alpha)$ *and* $q \in \mathscr{C}(\Sigma_\alpha)$.

Proof. Assume (i) does not hold. Since $\alpha \geqslant \beta$, we must have $\alpha_1 \geqslant \beta_1$. If $\alpha_1 > \beta_1$, then the number of *columns* in β is less than α_1. This implies that two of the entries of the first row of Σ_α must appear in Σ_β' in the same column. But as (i) fails to hold, this is impossible, and hence $\alpha_1 = \beta_1$. Since no two distinct elements of the first row of Σ_α lie in the same column of Σ_β', there exists a $q_1' \in \mathscr{C}(\Sigma_\beta')$ such that the first row of $q_1' \cdot \Sigma_\beta'$ has the same elements as the first row of Σ_α.

We may now ignore the first rows of Σ_α and $q_1' \cdot \Sigma_\beta'$ (which are the same up to order) and, using the definition of the lexicographical order, show similarly that $\alpha_2 = \beta_2$ and that there exists a $q_2' \in \mathscr{C}(\Sigma_\beta')$ such that Σ_α and $q_2'q_1' \cdot \Sigma_\beta'$ have the same first and second rows. Since α and β have the same number of cases, this process must give that $\alpha = \beta$ and that there exists a $q' \in \mathscr{C}(\Sigma_\beta')$ such that $q' \cdot \Sigma_\beta' = p \cdot \Sigma_\alpha$ for some $p \in \mathscr{R}(\Sigma_\alpha)$. Hence $\Sigma_\beta' = q'^{-1}p \cdot \Sigma_\alpha = p(p^{-1}q'^{-1}p) \cdot \Sigma_\alpha$. But $q'^{-1} \in \mathscr{C}(p \cdot \Sigma_\alpha) = p\mathscr{C}(\Sigma_\alpha)\, p^{-1}$ since $q'^{-1} \in \mathscr{C}(q' \cdot \Sigma_\beta') = q'\mathscr{C}(\Sigma_\beta')\, q'^{-1}$. Thus $p^{-1}q'^{-1}p = q \in \mathscr{C}(\Sigma_\alpha)$.

Corollary. *Suppose $s \in \mathscr{G}_f$ is not expressible as pq for any $p \in \mathscr{R}(\Sigma_\alpha)$ and $q \in \mathscr{C}(\Sigma_\alpha)$. Then there exist transpositions $u \in \mathscr{R}(\Sigma_\alpha)$ and $v \in \mathscr{C}(\Sigma_\alpha)$ such that $us = sv$.*

Proof. Since (ii) of the lemma fails for Σ_α and $\Sigma_\alpha' = s \cdot \Sigma_\alpha$ there exist $a, b \in \{1,..., f\}$ which are in one row of Σ_α and one column of $s \cdot \Sigma_\alpha$. Let $u \in \mathscr{R}(\Sigma_\alpha)$ be the transposition which interchanges a and b, and let $v = s^{-1}us$. Then clearly $v \in \mathscr{C}(\Sigma_\alpha)$, so the corollary follows.

We shall now determine the minimal left ideals of the group ring $A = \Omega[\mathscr{G}_f]$. Corresponding to the Young's diagram Σ_α, define $a_\alpha = \Sigma_{p \in \mathscr{R}(\Sigma_\alpha)}\, p$ and $b_\alpha = \Sigma_{q \in \mathscr{C}(\Sigma_\alpha)} \epsilon_q q$, where ϵ_q is the signature of q. The following are easy to verify:

(i) if $p \in \mathscr{R}(\Sigma_\alpha)$, then $pa_\alpha = a_\alpha p = a_\alpha$, and
(ii) if $q \in \mathscr{C}(\Sigma_\alpha)$, then $qb_\alpha = b_\alpha q = \epsilon_q b_\alpha$.

Lemma. *Let $c_\alpha = a_\alpha b_\alpha$. Then $c_\alpha \neq 0$.*

We can now prove our first main result.

Proposition 1. *Ac_α is a minimal left ideal in A.*

Two preliminary lemmas are needed.

Lemma 1. *In order that $x \in A$ be such that for every $p \in \mathscr{R}(\Sigma_\alpha)$ and every $q \in \mathscr{C}(\Sigma_\alpha)$ one has $pxq = \epsilon_q x$, it is necessary and sufficient that $x = \mu c_\alpha$ for some $\mu \in \Omega$.*

Lemma 2. *For $u \in A$, $a_\alpha u b_\alpha = \mu c_\alpha$ for some $\mu \in \Omega$.*

Proof. Lemma 2 follows from Lemma 1, for $x = a_\alpha u b_\alpha$ has the property of Lemma 1. To prove Lemma 1 first note that $pc_\alpha q = pa_\alpha b_\alpha q = \epsilon_q a_\alpha b_\alpha = \epsilon_q c_\alpha$. To prove the converse, let $x = \Sigma_{s \in \mathscr{G}_f} \lambda_s s$. The condition $pxq = \epsilon_q x$ implies that $\lambda_{pq} = \epsilon_q \lambda_e$. If $s \in \mathscr{G}_f$ is not of the form pq, then there exist transpositions $u \in \mathscr{R}(\Sigma_\alpha)$ and $v \in \mathscr{C}(\Sigma_\alpha)$ such that $usv^{-1} = s$. Applying the condition to uxv^{-1}, one obtains that $\lambda_{usv^{-1}} usv^{-1} = \epsilon_{v^{-1}} \lambda_s s$. But then $\lambda_s = -\lambda_s$, i.e., $\lambda_s = 0$. It follows that

$$\begin{aligned} x &= \Sigma_{p,q} \lambda_{pq} pq \\ &= \lambda_e \Sigma_{p,q} \epsilon_q pq \\ &= \lambda_e c_\alpha . \end{aligned}$$

Hence the necessity holds.

Proof of Proposition 1. Let $\mathscr{L} \neq \{0\}$ be a left ideal in A such that $\mathscr{L} \subset Ac_\alpha$. Then $c_\alpha \mathscr{L} \subset c_\alpha A c_\alpha \subset a_\alpha A b_\alpha \subset \Omega c_\alpha$ by Lemma 2. Hence $c_\alpha \mathscr{L}$ must be $\{0\}$ or Ωc_α. The first possibility cannot occur since $Ac_\alpha \mathscr{L} \supset \mathscr{L}^2 \neq \{0\}$ ($\mathscr{L}^2 \neq \{0\}$ as A is semi-simple). Thus $c_\alpha \mathscr{L} = \Omega c_\alpha$, and so $Ac_\alpha = Ac_\alpha \mathscr{L} \subset \mathscr{L}$, that is, $\mathscr{L} = Ac_\alpha$. Therefore Ac_α is minimal.

Proposition 2. *Suppose $\alpha \neq \beta$. Then the minimal left ideals Ac_α, corresponding to some Σ_α, and Ac_β, corresponding to some Σ_β, are not isomorphic (as A-modules).*

Proof. We may assume $\alpha > \beta$. We claim that $c_\alpha A c_\beta = \{0\}$, and hence $(Ac_\alpha)(Ac_\beta) = \{0\}$. Thus it is impossible that Ac_α and Ac_β are isomorphic. To verify the claim, it suffices to show $a_\alpha s b_\beta = 0$ for all $s \in \mathscr{G}_f$. First, $a_\alpha b_\beta = 0$. For since $\alpha > \beta$, there must be two distinct integers i, j in the same row of Σ_α and the same column of Σ_β. Let t be the transposition which interchanges i and j; $t \in \mathscr{R}(\Sigma_\alpha) \cap \mathscr{C}(\Sigma_\beta)$, so $a_\alpha t = a_\alpha$ and $t^{-1} b_\beta = -b_\beta$. It follows that $a_\alpha b_\beta = -a_\alpha b_\beta$, so that $a_\alpha b_\beta = 0$. Finally, $a_\alpha s b_\beta s^{-1} = 0$, since $s b_\beta s^{-1}$ is the b_β corresponding to $\Sigma_\beta' = s \cdot \Sigma_\beta$, and this gives the result.

Proposition 3. *Every minimal left ideal of A is isomorphic to some Ac_α.*

Proof. First of all, it is well known that a permutation $s \in \mathscr{G}_f$ decomposes uniquely into cycles $(i_1, \ldots, i_r)$ and that two permutations are conjugate if and only if the lengths of their cycles are equal. Thus the number of frames α is precisely the number of conjugate classes of $\mathscr{G}_f$. By Part IV of the Appendix, this is precisely the number of isomorphism classes of minimal left ideals in A, so the proof follows from Proposition 2.

We have shown that the number of isomorphism classes of minimal left ideals of $\Omega[\mathscr{G}_f]$ is equal to the number of frames with f cases. It follows from this that if c_α and c_α' correspond to the same frame (but to different diagrams), then $Ac_\alpha \simeq Ac_\alpha'$.

4. The Characters of $\mathbf{GL}\,(n; \Omega)$

We now turn to the second problem, that is, the determination of the 1-dimensional $A_n{}^f$-submodules of $E^{\otimes f}$. By § 3 and Part IV of the Appendix, every simple $A_n{}^f$-module is isomorphic to one of the form $c_\alpha \cdot E^{\otimes f}$, where α is a frame with f cases (although many of the $c_\alpha \cdot E^{\otimes f} = \{0\}$, due to the fact that the representation $\Omega[\mathscr{G}_f] \to \mathrm{End}(E^{\otimes f})$ is not always faithful). If c_α and c_α' (corresponding to different diagrams Σ_α and Σ_α') generate isomorphic minimal left ideals of $\Omega[\mathscr{G}_f]$ and if $c_\alpha \cdot E^{\otimes f}$ is simple, then $c_\alpha' \cdot E^{\otimes f}$ is simple and isomorphic to $c_\alpha \cdot E^{\otimes f}$. Hence the isomorphism class of $c_\alpha \cdot E^{\otimes f}$ depends only on α and not on Σ_α. We will denote this class by Σ_α and refer to it as a *space of irreducible tensors.*

We shall now determine the form of $c_\alpha(x_1 \otimes \cdots \otimes x_f)$ for $x_1, \ldots, x_f \in E$. Suppose the i-th column of α has length β_i. Then we may consider c_α for the Young's diagram

$$\begin{array}{lll} 1 & \beta_1 + 1 & \cdots f - \beta_r + 1 \\ 2 & \cdot & \qquad\quad \cdot \\ \cdot & \cdot & \qquad\quad f \\ \cdot & \beta_1 + \beta_2 & \\ \beta_1 & & \end{array}$$

Write $x_1 \otimes \cdots \otimes x_f = y_1 \otimes \cdots \otimes y_r$ where $y_1 = x_1 \otimes \cdots \otimes x_{\beta_1}$,

$$y_2 = x_{\beta_1+1} \otimes \cdots \otimes x_{\beta_1+\beta_2}, \ldots, y_r = x_{f-\beta_r+1} \otimes \cdots \otimes x_f .$$

Then

$$b_\alpha(y_1 \otimes \cdots \otimes y_r) = \Sigma_{q \in \mathscr{C}(\Sigma_\alpha)} \epsilon_q q(y_1 \otimes \cdots \otimes y_r).$$

Since $q \in \mathscr{C}(\Sigma_\alpha)$ can be written $q = q_1 \cdots q_r$ where q_i acts only on the i-th column of Σ_α, we have

$$\begin{aligned} b_\alpha(y_1 \otimes \cdots \otimes y_r) &= \Sigma_{q_1, \ldots, q_r} \epsilon_{q_1} \cdots \epsilon_{q_r} q_1 y_1 \otimes \cdots \otimes q_r y_r \\ &= \Sigma_{q_1} \epsilon_{q_1} q_1 y_1 \otimes \cdots \otimes \Sigma_{q_r} \epsilon_{q_r} q_r y_r \\ &= \underline{a}(y_1) \otimes \cdots \otimes \underline{a}(y_r). \end{aligned}$$

$\underline{a}(y_i)$ is called the *antisymmetrization* of the β_i-tensor y_i. In the terminology of exterior algebra, $\underline{a}(y_i)$ can be identified with an m-vector ($m = \beta_i$), since Ω has characteristic 0. Since for $m > n = \dim E$ an m-vector is 0, we have the first result.

Proposition. *$c_\alpha \cdot E^{\otimes f} = \{0\}$ if (and only if) the first column of α has length greater than n.*

Example. If α has only one row, then $c_\alpha = a_\alpha$, and $c_\alpha(x_1 \otimes \cdots \otimes x_f) = \Sigma_{s\in\mathscr{G}_f} s(x_1 \otimes \cdots \otimes x_f)$. Thus c_α is the symmetrization operator $\underline{s}$. If α has only one column, the $c_\alpha = b_\alpha$, and $c_\alpha(x_1 \otimes \cdots \otimes x_f) = \underline{a}(x_1 \otimes \cdots \otimes x_f)$. In the case $f = 2$ (since there are only two frames) we obtain a classical result; namely, every tensor of order two decomposes uniquely as the sum of a symmetric tensor and a skew-symmetric tensor. In the case $f = 3$ there is an additional frame and so other irreducible tensors must occur.

Tensors in Σ_α are said to have *signature* $\alpha = (\alpha_1, \ldots, \alpha_r)$. In particular, an alternating tensor has signature $(1, 1, \ldots, 1)$ and a symmetric tensor has signature $(f, 0, \ldots, 0)$.

Proposition. *In order that Σ_α be 1-dimensional, it is necessary and sufficient that each column of α have length $n = \dim E$.*

Proof. By (1), the sufficiency is clear, since the n-th exterior power of E is 1-dimensional. Suppose now that $n = \beta_1 = \cdots = \beta_{h-1} > \beta_h = \cdots = \beta_m \geqslant \cdots$ for some $h \leqslant r$. Choose a basis $e_1, \ldots, e_n$ of E and define $y_i = e_1 \otimes \cdots \otimes e_{\beta_i}$. Then $c_\alpha(y_1 \otimes \cdots \otimes y_r) = k\underline{a}(y_1) \otimes \cdots \otimes \underline{a}(y_r)$, where k is the order of $\mathscr{R}(\Sigma_\alpha)$. By exterior algebra, $c_\alpha(y_1 \otimes \cdots \otimes y_r) \neq 0$.

Now let m' be an integer such that $\beta_h < m' \leqslant n$ and let s be the transposition which interchanges β_h and m'. Define

$$y_i' = \begin{cases} s(y_i) & \text{if} \quad i < h \\ e_1 \otimes \cdots \otimes e_{\beta_h - 1} \otimes e_{m'} & \text{if} \quad h \leqslant i \leqslant m \\ y_i & \text{if} \quad i > m. \end{cases}$$

Let $z = x_1 \otimes \cdots \otimes x_f$ and let $p \in \mathscr{G}_f$. Then $p \cdot z = x_1' \otimes \cdots \otimes x_f'$ where $x_i' = x_{p^{-1}(i)}$. Define

$$y_1^p = x_1' \otimes \cdots \otimes x_{\beta_1}', \quad y_2^p = x_{\beta_1+1}' \otimes \cdots \otimes x_{\beta_1+\beta_2}',$$

and so forth. Then

$$c_\alpha(x_1 \otimes \cdots \otimes x_f) = \Sigma_{p\in\mathscr{R}(\Sigma_\alpha)} \underline{a}(y_1^p) \otimes \cdots \otimes \underline{a}(y_r^p). \tag{1}$$

Then

$$\begin{aligned} c_\alpha(y_1' \otimes \cdots \otimes y_r') &= k\underline{a}(y_1') \otimes \cdots \otimes \underline{a}(y_r') \\ &= \pm k\underline{a}(y_1) \otimes \cdots \otimes \underline{a}(y_{h-1}) \otimes \underline{a}(y_h') \otimes \cdots \otimes \underline{a}(y_r'). \end{aligned}$$

By examination of the basis of

$$\underbrace{\underline{a}(E^{\otimes n}) \otimes \cdots \otimes \underline{a}(E^{\otimes n})}_{h-1} \otimes \underbrace{\underline{a}(E^{\otimes \beta_h}) \otimes \cdots \otimes \underline{a}(E^{\otimes \beta_h})}_{m-h} \otimes \cdots$$

where $\underline{a}(E^{\otimes i})$ denotes the i-th exterior power of E, one sees that $c_\alpha(y_1 \otimes \cdots \otimes y_r)$ and $c_\alpha(y_1' \otimes \cdots \otimes y_r')$ must be linearly independent. Hence if $\beta_h < n$, $c_\alpha \cdot E^{\otimes f}$ is at least of dimension two, which contradicts our hypothesis.

Now let $\underline{X} \in \mathbf{GL}(n; \Omega)$ and let α be the rectangular frame $(g, g, \ldots, g)$ with n rows. Then $f = gn$, and

$$\begin{aligned} \underline{X}^{\otimes f} \cdot (c_\alpha(x_1 \otimes \cdots \otimes x_f)) &= \sum_{p \in \mathscr{R}(\Sigma_\alpha)} \underline{a}(\underline{X}^{\otimes n} y_1{}^p) \otimes \cdots \otimes \underline{a}(\underline{X}^{\otimes n} y_g{}^p) \\ &= \det(\underline{X})^g \sum_{p \in \mathscr{R}(\Sigma_\alpha)} \underline{a}(y_1{}^p) \otimes \cdots \otimes \underline{a}(y_g{}^p) \\ &= \det(\underline{X})^g \, c_\alpha(x_1 \otimes \cdots \otimes x_f). \end{aligned}$$

Therefore, $\underline{X}^{\otimes f} \mid c_\alpha \cdot E^{\otimes f}$ is multiplication by $\det(\underline{X})^g$. Certainly any two $A_n{}^f$-modules of dimension one (over Ω) must be isomorphic as modules. Thus any 1-dimensional (hence homogeneous) polynomial representation F of $\mathbf{GL}(n; \Omega)$ of degree f defines two isomorphic $A_n{}^f$-modules $c_\alpha \cdot E^{\otimes f} \cong \Omega$. Since $c_\alpha(y \otimes \cdots \otimes y)$, $y = e_1 \otimes \cdots \otimes e_n$, generates $c_\alpha \cdot E^{\otimes f}$, it follows that $F(\underline{X}) = \det(\underline{X})^g$ by letting the isomorphism be

$$c_\alpha(y \otimes \cdots \otimes y) \to 1.$$

Theorem. *Every rational abelian character of* $\mathbf{GL}(n; \Omega)$ *is of the form* $\underline{X} \to \det(\underline{X})^g$ *for some integer* g.

5. Multilinear Invariants of $\mathbf{GL}(n; \Omega)$

We are now able to determine the multilinear simultaneous relative invariants of $\Gamma = \mathbf{GL}(n; \Omega)$ of an arbitrary number of vectors in $E = \Omega^n$. Let $v : E^{\otimes f} \to \Omega$ be a linear (simultaneous) relative invariant and suppose $c_\alpha \cdot E^{\otimes f}$ is an irreducible Γ-invariant subspace of $E^{\otimes f}$ corresponding to a Young's frame α. Letting $v_\alpha = v \mid c_\alpha \cdot E^{\otimes f}$, one notes that, by Schur's Lemma, v_α is either 0 or a bijection. Hence $v_\alpha = 0$ unless $c_\alpha \cdot E^{\otimes f}$ is 1-dimensional, and since $E^{\otimes f}$ is a direct sum of irreducible subspaces, non-trivial linear relative invariants on $E^{\otimes f}$ exist only if $f = gn$ for some integer g. Another way to see this is to prove that v is associated with a polynomial character on E of degree f and apply the last section.

Let $x_1, \ldots, x_n \in E$ be vectors with $x_i = \Sigma \xi_i^j e_j$. We define the *bracket* $[x_1 \cdots x_n]$ of $x_1, \ldots, x_n$ to be $\det(\xi_i^j)$. The bracket is a multilinear invariant of weight one. (Since the weight of any relative invariant is $\det(\underline{X})^g$ for some integer g, we simply refer to g as the weight of the invariant.) The next theorem is the first version of the *first main theorem on invariants.*

Theorem. *Multilinear invariants of* $\mathbf{GL}(n; \Omega)$ *of f vectors in* $E = \Omega^n$ *exist only if f is a multiple of n, say $f = gn$. Each such is a linear combination of invariants of the form*

$$[x_{i_1} \cdots x_{i_n}][x_{i_{n+1}} \cdots x_{i_{2n}}] \cdots [x_{i_{(f-n+1)}} \cdots x_{i_f}] \tag{1}$$

where $(i_1 \cdots i_f)$ is an arbitrary permutation of $\mathscr{G}_f$. The weight of each such invariant is g.

Proof. The first assertion follows from Proposition 4 of Chapter 1 and the above remarks. Let u be a multilinear invariant of $f = gn$ vectors, and replace u by a linear invariant v on $E^{\otimes f}$. Assume $\alpha = (g, g, \ldots, g)$ is the unique frame such that E_α has dimension one. If $c_\alpha^{(1)} \cdot E^{\otimes f}, \ldots, c_\alpha^{(r)} \cdot E^{\otimes f}$ are the 1-dimensional irreducible subspaces that occur in some direct sum decomposition of $E^{\otimes f}$, then

$$v(x_1 \otimes \cdots \otimes x_f) = \sum_{i=1}^{r} v(c_\alpha^{(i)}(x_1 \otimes \cdots \otimes x_f)).$$

Since $\underline{a}(x_1 \otimes \cdots \otimes x_n) = [x_1 \cdots x_n]\, \underline{a}(e_1 \otimes \cdots \otimes e_n)$, the computation of the last section implies that

$$v(c_\alpha(x_1 \otimes \cdots \otimes x_f)) = \sum_{p \in \mathscr{R}(\Sigma_\alpha)} \lambda_p [x_{p(1)} \cdots x_{p(n)}] \cdots [x_{p(f-n+1)} \cdots x_{p(f)}]$$

where $\lambda_p \in \Omega$. This completes the proof.

Absolute multilinear invariants of $\mathbf{GL}(n; \Omega)$ cannot, therefore, exist.

Rational absolute invariants can easily be constructed, however. For example,

$$[x_1x_2][x_3x_4]/[x_1x_3][x_2x_4],$$

the cross ratio of four vectors in the plane, is an absolute invariant.

We now consider some applications of this theorem.

(1) *Multilinear invariants of several tensors*

Assume that for $i = 1,\ldots, r$ one has tensor spaces $M_i = E^{\otimes f_i}$ and a multilinear invariant u: $M_1 \times \cdots \times M_r \to \Omega$. Let v: $M_1 \otimes \cdots \otimes M_r \to \Omega$ be the associated linear invariant, and consider for each $i = 1,\ldots, r$ decomposable tensors $z_i = \overset{i}{x}_1 \otimes \overset{i}{x}_2 \otimes \cdots \otimes \overset{i}{x}_{f_i} \in M_i$. By the universal property of the tensor product the mapping $w(\overset{1}{x}_1,\ldots, \overset{1}{x}_{f_1}, \overset{2}{x}_1,\ldots, \overset{r}{x}_{f_r}) = v(z_1 \otimes \cdots \otimes z_r)$ must be a multilinear invariant of $f = \Sigma f_i$ vectors in E, so $f = gn$ and w must have the form (1). w completely determines v, and, by what is known as the *first process of restitution*, the form of v may be deduced from w.

We now describe this process. Suppose that, with respect to a basis $e_1,\ldots, e_n$ of E, $\overset{j}{x}_k$ has components $(\underset{j}{\xi}^{k1},\ldots, \underset{j}{\xi}^{kn})$. Then setting for simplicity $f_j = s$,

$$z_j = \Sigma \underset{j}{\xi}^{1l_1} \cdot \underset{j}{\xi}^{2l_2} \cdots \underset{j}{\xi}^{sl_s} e_{l_1} \otimes \cdots \otimes e_{l_s}$$

where the summation is taken over all indices $l_1,\ldots, l_s$ with $1 \leqslant l_1,\ldots, l_s \leqslant n$. For an arbitrary tensor $y_j = \Sigma \underset{j}{\zeta}^{l_1\cdots l_s} e_{l_1} \otimes \cdots \otimes e_{l_s}$ of M_j, $v(y_1 \otimes \cdots \otimes y_r)$ is obtained by replacing, for every j, each component $\underset{j}{\xi}^{1l_1} \cdots \underset{j}{\xi}^{sl_s}$ of z_j by the corresponding component $\underset{j}{\zeta}^{l_1\cdots l_s}$ of y_j in the expression for $v(z_1 \otimes \cdots \otimes z_r)$.

For example, suppose $n = 3$, $M_1 = E \otimes E$, and $M_2 = E$. Consider the multilinear invariant on $E \times E \times E$ given by $[\overset{1}{x}_1 \overset{1}{x}_2 \overset{2}{x}]$. Set

$$\overset{1}{x}_1 = \underset{1}{\xi}^{11}e_1 + \underset{1}{\xi}^{12}e_2 + \underset{1}{\xi}^{13}e_3, \overset{1}{x}_2 = \underset{1}{\xi}^{21}e_1 + \underset{1}{\xi}^{22}e_2 + \underset{1}{\xi}^{23}e_3,$$

and

$$\overset{2}{x} = \eta^1 e_1 + \eta^2 e_2 + \eta^3 e_3.$$

Now

$$[\overset{1}{x}_1\overset{1}{x}_2\overset{2}{x}] = \eta^1(\underset{1}{\xi}^{12}\underset{1}{\xi}^{23} - \underset{1}{\xi}^{13}\underset{1}{\xi}^{22}) + \eta^2(\underset{1}{\xi}^{11}\underset{1}{\xi}^{23} - \underset{1}{\xi}^{13}\underset{1}{\xi}^{21}) + \eta^3(\underset{1}{\xi}^{11}\underset{1}{\xi}^{22} - \underset{1}{\xi}^{12}\underset{1}{\xi}^{21}).$$

If $y \in E \otimes E$ has components ζ^{ij}, one replaces $\underset{1}{\xi}^{1i}\underset{1}{\xi}^{2j}$ by ζ^{ij}, getting an invariant

$$v(y, \overset{2}{x}) = \eta^1(\zeta^{23} - \zeta^{32}) + \eta^2(\zeta^{13} - \zeta^{31}) + \eta^3(\zeta^{12} - \zeta^{21})$$

on $M_1 \times M_2$. The invariant $[\overset{1}{x}_1\overset{1}{x}_2\overset{2}{x}]$ is called the *symbolic expression* of the invariant v.

(2) *Homogeneous invariants of several tensors*

Suppose f is a homogeneous polynomial invariant of three tensors $x \in E^{\otimes p}$, $y \in E^{\otimes q}$, and $z \in E^{\otimes r}$ of degrees h in the components of x, h' in the components of y, and h'' in the components of z. Polarizing f, one obtains a multilinear invariant u in the tensors $\overset{1}{x},\dots,\overset{h}{x} \in E^{\otimes p}$, $\overset{1}{y},\dots,\overset{h'}{y} \in E^{\otimes q}$, and $\overset{1}{z},\dots,\overset{h''}{z} \in E^{\otimes r}$. Thus the degrees h, h', and h'' must satisfy a relation $hp + h'q + h''r = gn$. By the last example, we are led to consider multilinear invariants u of gn vectors $\overset{11}{x},\dots,\overset{1p}{x}, \overset{21}{x},\dots,\overset{h1}{x},\dots,\overset{hp}{x}, \overset{11}{y},\dots,\overset{h'q}{y}, \overset{11}{z},\dots,\overset{h''r}{z} \in E$. To obtain f, one must restitute the symbolic expression of u twice.

Take, for example, $n = 2$, and let us compute all homogeneous invariants of degree two on $E \otimes E$. Here $p = 2$, $h = 2$, $n = 2$, and so $g = 2$. Write $x = \Sigma\zeta^{ij}e_i \otimes e_j$ and introduce $\overset{1}{x} = \Sigma\underset{1}{\zeta}^{ij}e_i \otimes e_j$ and $\overset{2}{x} = \Sigma\underset{2}{\zeta}^{ij}e_i \otimes e_j$. In turn, introduce $\overset{11}{x}, \overset{12}{x}, \overset{21}{x}, \overset{22}{x} \in E$ with $\overset{rs}{x} = \underset{r}{\xi}^{s1}e_1 + \underset{r}{\xi}^{s2}e_2$. Consider first the invariant

$$[\overset{11}{x}\,\overset{22}{x}][\overset{12}{x}\,\overset{21}{x}] = (\underset{1}{\xi}^{11}\underset{2}{\xi}^{22} - \underset{1}{\xi}^{12}\underset{2}{\xi}^{21})(\underset{1}{\xi}^{21}\underset{2}{\xi}^{12} - \underset{1}{\xi}^{22}\underset{2}{\xi}^{11}).$$

The first restitution gives the invariant

$$\underset{1}{\zeta}^{11}\underset{2}{\zeta}^{22} - \underset{1}{\zeta}^{12}\underset{2}{\zeta}^{12} - \underset{1}{\zeta}^{21}\underset{2}{\zeta}^{21} + \underset{1}{\zeta}^{22}\underset{2}{\zeta}^{11}$$

of $\overset{1}{x}$ and $\overset{2}{x}$. The final restitution (which amounts to erasing the lower indices) gives the *homogeneous* invariant $2\zeta^{11}\zeta^{22} - (\zeta^{12})^2 - (\zeta^{21})^2$ on $E \otimes E$. Starting with the symbolic expression $[\overset{11}{x}\,\overset{12}{x}][\overset{21}{x}\,\overset{12}{x}]$ one obtains the invariant $(\zeta^{12} - \zeta^{21})^2$, and starting with $[\overset{11}{x}\,\overset{21}{x}][\overset{12}{x}\,\overset{22}{x}]$ gives $2(\zeta^{11}\zeta^{22} - \zeta^{12}\zeta^{21})$. Thus every homogeneous invariant on $E \otimes E$ of degree two is a linear combination of these three invariants.

(3) *Multilinear invariants of irreducible tensors*

We consider now the case where f is a bilinear invariant of $x \in c_\alpha \cdot E^{\otimes p}$ and $y \in c_\beta' \cdot E^{\otimes q}$. Thus $f(\sigma \cdot x, \sigma \cdot y) = (\det \sigma)^g f(x, y)$ for $\sigma \in \mathbf{GL}(n; \Omega)$. By the Appendix $c_\alpha{}^2 = \mu_\alpha c_\alpha$ and $c_\beta'^2 = \mu_\beta' c_\beta'$ for some non-zero scalars μ_α and $\mu_\beta' \in \Omega$. For $u \in E^{\otimes p}$ and $v \in E^{\otimes q}$ define $\varphi(u, v) = f(c_\alpha u, c_\beta' v)$. Then, by the commutation property,

$$\begin{aligned}\varphi(\sigma \cdot u, \sigma \cdot v) &= f(c_\alpha \sigma \cdot u, c_\beta' \sigma \cdot v)\\ &= f(\sigma \cdot c_\alpha u, \sigma \cdot c_\beta' v)\\ &= (\det \sigma)^g f(c_\alpha u, c_\beta' v)\\ &= (\det \sigma)^g \varphi(u, v)\end{aligned}$$

so φ is a bilinear invariant on $E^{\otimes p} \times E^{\otimes q}$ such that $\varphi(c_\alpha u, c_\beta' v) = \mu_\alpha \mu_\beta' f(c_\alpha u, c_\beta' v)$. Hence we are led to consider an earlier situation.

6. Invariants of Antisymmetric Tensors

As a special case of the last paragraph, we now begin a study of multilinear invariants of antisymmetric tensors, which will lead to a more general formulation of the main theorem. Recall that a multilinear map $f : E \times E \times \cdots \times E \to F$ of p vectors is called alternating if $f(\overset{\sigma(1)}{x}, \ldots, \overset{\sigma(p)}{x}) = \epsilon_\sigma f(\overset{1}{x}, \ldots, \overset{p}{x})$ for every $\sigma \in \mathscr{G}_p$ and that, by the universal property of the p-th exterior power $\Lambda^p E$, f defines a homomorphism $f' : \Lambda^p E \to F$ such that $f(\overset{1}{x}, \ldots, \overset{p}{x}) = f'(\overset{1}{x} \wedge \cdots \wedge \overset{p}{x})$. Now $\mathbf{GL}(E)$ acts naturally on $\Lambda^p E$ by the tensor representation $U \to \Lambda^p U$, where $\Lambda^p U$ is the isomorphism of $\Lambda^p E$ induced by the alternating map

$$(\overset{1}{x}, \ldots, \overset{p}{x}) \to U(\overset{1}{x}) \wedge \cdots \wedge U(\overset{p}{x}).$$

Suppose that $\Gamma \subset \mathbf{GL}(n; \Omega)$ and that g is a multilinear invariant of an antisymmetric tensor u of order $p < n$ (equivalently, a p-vector u') and of a certain number of vectors $\overset{1}{y}, \ldots, \overset{q}{y} \in E$. One can associate with g a multilinear invariant f given by $f(\overset{1}{x}, \ldots, \overset{p}{x}, \overset{1}{y}, \ldots, \overset{q}{y}) = g(\overset{1}{x} \wedge \cdots \wedge \overset{p}{x}, \overset{1}{y}, \ldots, \overset{q}{y})$ for $\overset{1}{x}, \ldots, \overset{p}{x} \in E$. Clearly, f is alternating in $\overset{1}{x}, \ldots, \overset{p}{x}$. Conversely, given a multilinear invariant $f(\overset{1}{x}, \ldots, \overset{p}{x}, \overset{1}{y}, \ldots, \overset{q}{y})$, then

$$g(\overset{1}{x} \wedge \cdots \wedge \overset{p}{x}, \overset{1}{y}, \ldots, \overset{q}{y}) = (1/p!) \sum_{\sigma \in \mathscr{G}_p} \epsilon_\sigma f(\overset{\sigma(1)}{x}, \ldots, \overset{\sigma(p)}{x}, \overset{1}{y}, \ldots, \overset{q}{y})$$

determines a multilinear invariant of a p-vector z and $\overset{1}{y}, \ldots, \overset{q}{y}$.

When Γ is $\mathbf{GL}(n; \Omega)$, f is a product of determinants in the $\overset{i}{x}$ and $\overset{j}{y}$. A particularly simple case is when each of the $\overset{i}{x}$ appear in the same bracket. Since, for $\sigma \in \mathcal{G}_p$, $\epsilon_\sigma[\overset{\sigma(1)}{x} \cdots \overset{\sigma(p)}{x}\overset{1}{y} \cdots \overset{r}{y}] = [\overset{1}{x} \cdots \overset{p}{x}\overset{1}{y} \cdots \overset{r}{y}]$ $(r = n - p)$, $g(u, \overset{1}{y}, \ldots, \overset{q}{y})$ is obtained (up to a factor) by developing $[\overset{1}{x} \cdots \overset{p}{x}\overset{1}{y} \cdots \overset{r}{y}]$ along the first p lines using Laplace's rule and replacing by $\pi^{i_1 \cdots i_p}$ (the $(i_1 \cdots i_p)$-th component of u) the minor formed by these lines and the columns with indices $i_1, \ldots, i_p$, where $i_1 < i_2 < \cdots < i_p$.

We will now show that the general case reduces to the above case with the aid of a general identity of invariant theory. Given a polynomial $\varphi(z_1, \ldots, z_m)$, in the components of the vectors $z_1, \ldots, z_m \in E$, we define, for $1 \leqslant k \leqslant m$,

$$\varphi_k(z_1, \ldots, z_m) = \sum_{\sigma \in \mathcal{G}_k} \epsilon_\sigma \varphi(z_{\sigma(1)}, \ldots, z_{\sigma(k)}, z_{k+1}, \ldots, z_m). \tag{1}$$

It is possible to calculate φ_k by recursion on k with the aid of the following identity:

$$\varphi_{k+1}(z_1, \ldots, z_m) = \varphi_k(z_1, \ldots, z_m) + \sum_{j=1}^{k} (-1)^{k-j-1} \varphi_k(z_1, \ldots, \hat{z}_j, \ldots, z_{k+1}, z_j, z_{k+2}, \ldots, z_m) \tag{2}$$

(hatted terms are omitted). To establish (2), one considers, for each j $(1 \leqslant j \leqslant k + 1)$, all the terms in the expression of φ_{k+1} for which $\sigma(k + 1) = j$. Let $\rho : \{1, \ldots, k\} \to \{1, \ldots, \hat{j}, \ldots, k + 1\}$ be defined by $\rho(i) = i$ if $i < j$ and $\rho(i) = i + 1$ if $i \geqslant j$. For every $\sigma \in \mathcal{G}_{k+1}$ such that $\sigma(k + 1) = j$, define an element π of $\mathcal{G}_k$ by $\pi = \rho^{-1}\sigma$. The mapping $\sigma \to \pi$ is a bijection of these σ's onto $\mathcal{G}_k$, and $\epsilon_\sigma = (-1)^{k-j+1}\epsilon_\pi$. Setting $z_i' = z_{\rho(i)}$ if $i \leqslant k$ we see that

$$\begin{aligned}
&\sum_{\sigma(k+1)=j} \epsilon_\sigma \varphi(z_{\sigma(1)}, \ldots, z_{\sigma(k)}, z_j, z_{k+2}, \ldots, z_m) \\
&= \Sigma_\sigma \epsilon_\sigma \varphi(z'_{\pi(1)}, \ldots, z'_{\pi(k)}, z_j, z_{k+2}, \ldots, z_m) \qquad (\rho\pi = \sigma) \\
&= (-1)^{k-j-1}\, \Sigma_\pi \epsilon_\pi \varphi(z'_{\pi(1)}, \ldots, z'_{\pi(k)}, z_j, \ldots, z_m) \\
&= (-1)^{k-j-1}\, \varphi_k(z_1', \ldots, z_k', z_j, \ldots, z_m) \\
&= (-1)^{k-j-1}\, \varphi_k(z_1, \ldots, \hat{z}_j, \ldots, z_{k+1}, z_j, z_{k+2}, \ldots, z_m).
\end{aligned}$$

Finally, summing over j gives the desired identity (2).

Now, consider a multilinear invariant f of the form $f(\overset{1}{x},\dots,\overset{p}{x},\overset{1}{y},\dots,\overset{q}{y}) = [\overset{1}{x}\dots\overset{k}{x}\overset{1}{y}\dots\overset{r}{y}]u(\overset{k+1}{x})$, where $k < p$, $p \leqslant n$, $r = n - k$, and $u(\overset{k+1}{x})$ is a linear form in $\overset{k+1}{x}$ whose coefficients depend on $\overset{k+2}{x},\dots,\overset{p}{x},\overset{r+1}{y},\dots,\overset{q}{y}$. Since $f_k = k!f$, (2) gives

$$f_{k+1}(\overset{1}{x},\dots,\overset{p}{x},\overset{1}{y},\dots,\overset{q}{y}) = k!\,[\overset{1}{x}\cdots\overset{k}{x}\overset{1}{y}\cdots\overset{r}{y}]\,u(\overset{k+1}{x})$$

$$+\,k!\sum_{j=1}^{k}(-1)^{k-j-1}\,[\overset{1}{x}\cdots\overset{j-1}{x}\;\overset{j+1}{x}\cdots\overset{k+1}{x}\;\overset{1}{y}\cdots\overset{r}{y}]\,u(\overset{j}{x}).$$

But as u is a linear form,

$$\begin{vmatrix} u(\overset{1}{x}) & \xi^1_1 & \cdots & \xi^n_1 \\ \vdots & & & \\ u(\overset{k+1}{x}) & \xi^1_{k+1} & \cdots & \xi^n_{k+1} \\ u(\overset{1}{y}) & \overset{1}{\underset{1}{\eta}} & \cdots & \overset{n}{\underset{1}{\eta}} \\ \vdots & \vdots & & \vdots \\ u(\overset{r}{y}) & \overset{1}{\underset{r}{\eta}} & & \overset{n}{\underset{r}{\eta}} \end{vmatrix} = 0.$$

Expanding this identity along the first column, we obtain

$$\sum_{j=1}^{k+1}(-1)^j\,[\overset{1}{x}\cdots\overset{j-1}{x}\;\overset{j+1}{x}\cdots\overset{k+1}{x}\;\overset{1}{y}\cdots\overset{r}{y}]\,u(\overset{j}{x})$$

$$= (-1)^k\sum_{h=1}^{r}(-1)^h\,[\overset{1}{x}\cdots\overset{k+1}{x}\;\overset{1}{y}\cdots\overset{h-1}{y}\;\overset{h+1}{y}\cdots\overset{r}{y}]\,u(\overset{h}{y}). \tag{3}$$

But by (2) the term on the left hand side of (3) is precisely $(-1)^{k+1}(k!)^{-1}f_{k+1}$, and hence repeated application of this argument shows that f_p is a sum of invariants, in all of which $\overset{1}{x},\dots,\overset{p}{x}$ figure in the same determinant.

As an application, if an invariant $f(\overset{1}{x},\dots,\overset{p}{x},\overset{1}{y},\dots,\overset{q}{y})$ is alternating in $\overset{1}{x},\dots,\overset{p}{x}$ then $f_p = f$ up to a factor $p!$ and thus this result holds for f.

7. Invariants of Mixed Tensors

Let E^* denote the dual space $\mathrm{Hom}(E, \Omega)$ of E. $\mathbf{GL}(E)$ acts on E^* as follows: for $U \in \mathbf{GL}(E)$, define $\hat{U} \in \mathbf{GL}(E^*)$ by the relation $\langle x, \hat{U}(x^*)\rangle =$

$\langle U^{-1}(x), x^*\rangle$ for all $x \in E$ and $x^* \in E^*$. The representation of $\mathbf{GL}(E)$ in $\mathbf{GL}(E^*)$ given by $U \to \hat{U}$ is irreducible. Let $T_q{}^p(E) = E^{\otimes p} \otimes (E^*)^{\otimes q}$ denote the space of mixed tensors of order (p, q), and let $\mathbf{GL}(E)$ act on $T_q{}^p(E)$ by means of the tensorial representation $U \to U^{\otimes p} \otimes \hat{U}^{\otimes q}$. Elements of $T_0{}^p(E)$ are called *contravariant tensors of order* p, and elements of $T_q{}^0(E)$ are called *covariant tensors of order* q. Given $x^* \in E^*$, one can define uniquely an $(n-1)$-vector $\varphi(x^*) \in \Lambda^{n-1}E$ characterized by the relation $x \wedge \varphi(x^*) = \langle x, x^*\rangle\, e_1 \wedge \cdots \wedge e_n$ for all $x \in E$, where $e_1 ,..., e_n$ is a basis of E. If x^* has components $(\xi_1 ,..., \xi_n)$ (with respect to the dual basis of E^*), then

$$\varphi(x^*) = \sum_{i=1}^{n} (-1)^{i-1}\, \xi_i e_1 \wedge \cdots \wedge \hat{e}_i \wedge \cdots \wedge e_n .$$

In other words, the i-th component $\pi^{1\cdots \hat{i}\cdots n}$ of $\varphi(x^*)$ is $(-1)^{i-1}\xi_i$. $\varphi : E^* \to \Lambda^{n-1}E$ is thus an isomorphism, and, by a simple computation, $U \cdot (\varphi(x^*)) = \det(U)\, \varphi(U \cdot x^*)$ for all $U \in \mathbf{GL}(E)$. Therefore, the representations $U \to \Lambda^{n-1}U$ and $U \to \hat{U}$ are equivalent.

Suppose $\Gamma \subset \mathbf{GL}(n; \Omega)$ and that f is a multilinear invariant of p contravariant vectors $\overset{1}{x},..., \overset{p}{x} \in E$ and q covariant vectors $y_1{}',..., y_q{}' \in E^*$. To determine f we can associate with f a multilinear invariant of $\overset{1}{x},...,\overset{p}{x} \in E$ and $\overset{1}{z},..., \overset{q}{z} \in \Lambda^{n-1}E$ by setting

$$h(\overset{1}{x},..., \overset{p}{x}, \overset{1}{z},..., \overset{q}{z}) = f(\overset{1}{x},..., \overset{p}{x}, \varphi^{-1}(\overset{1}{z}),..., \varphi^{-1}(\overset{q}{z})).$$

One can now apply the methods of the last paragraph. In order to study h we must introduce, for each index j $(1 \leqslant j \leqslant q)$, $n-1$ contravariant vectors $\overset{jk}{z}$ $(1 \leqslant k \leqslant n-1)$ and consider the multilinear invariant h' of the $\overset{i}{x}$ and $\overset{jk}{z}$ obtained from h by the general method.

When $\Gamma = \mathbf{GL}(n; \Omega)$, h' is a linear combination of products of determinants. Since h' is alternating in $\overset{11}{z},..., \overset{1,n-1}{z}$, we may suppose that $h' = [\overset{11}{z} \cdots \overset{1,n-1}{z}\; w]\, h''$, where w is one of the remaining vectors, and h'' does not contain any $\overset{1k}{z}$. After restituting $\overset{11}{z},..., \overset{1,n-1}{z}$ to $\overset{1}{z}$ and then to $y_1{}' = \varphi^{-1}(\overset{1}{z})$, $[\overset{11}{z} \cdots \overset{1,n-1}{z}\; w]$ becomes $\langle w, y_1{}'\rangle$. One can continue the restitution in this manner, as long as there remain brackets containing at least one $\overset{jk}{z}$, getting a linear combination of products whose factors are of the following types:

(i) contravariant brackets $[\overset{i_1}{x} \cdots \overset{i_n}{x}]$,

(ii) scalar products $\langle \overset{i}{x}, y_j' \rangle$,

(iii) products of the form $\langle \overset{j1}{z}, y_{i_1}' \rangle \langle \overset{j2}{z}, y_{i_2}' \rangle \cdots \langle \overset{j,n-1}{z}, y_{i_{n-1}}' \rangle$.

Antisymmetrizing (iii) gives

$$\sum_{\sigma \in \mathscr{S}_{n-1}} \epsilon_\sigma \langle \overset{j\sigma(1)}{z}, y_{i_1}' \rangle \cdots \langle \overset{j\sigma(n-1)}{z}, y_{i_{n-1}}' \rangle$$

$$= \det(\langle \overset{jk}{z}, y_{i_r}' \rangle)$$

$$= \langle \overset{j1}{z} \wedge \cdots \wedge \overset{j,n-1}{z}, y_{i_1}' \wedge \cdots \wedge y_{i_{n-1}}' \rangle.$$

Suppose that $y' \in E^*$ has components $(\xi_1, \ldots, \xi_n)$ and that $z = \varphi(y')$. Then

$$\langle z, y_{i_1}' \wedge \cdots \wedge y_{i_{n-1}}' \rangle$$

$$= \sum_{i=1}^{n} (-1)^{i-1} \xi_i \langle e_1 \wedge \cdots \wedge \hat{e}_i \wedge \cdots \wedge e_n, y_{i_1}' \wedge \cdots \wedge y_{i_{n-1}}' \rangle$$

$$= \sum_{i=1}^{n} (-1)^{i-1} \xi_i \det(\langle e_j, y_{i_k}' \rangle) \quad (j \neq i)$$

$$= [y' y_{i_1}' \cdots y_{i_{n-1}}'].$$

Hence the restitution of $\langle \overset{j1}{z} \wedge \cdots \wedge \overset{j,n-1}{z}, y_{i_1}' \wedge \cdots \wedge y_{i_{n-1}}' \rangle$ gives the contravariant bracket $[y_j' y_{i_1}' \cdots y_{i_{n-1}}']$ where $y_j' = \varphi^{-1}(\overset{i}{z})$. We summarize these results in the second version of the first main theorem on invariants.

Theorem. *Every multilinear relative invariant of* $\mathbf{GL}(n; \Omega)$ *of p vectors in E and q vectors of E^* is a linear combination of products of invariants of three types:*

(i) *contravariant brackets* $[\overset{1}{x} \cdots \overset{n}{x}]$ *of weight* 1,

(ii) *covariant brackets* $[y_1' \cdots y_n']$ *of weight* -1, *and*

(iii) *scalar products* $\langle x, y' \rangle$ *of weight* 0.

In particular, multilinear invariants can exist only if $p - q$ is a multiple of n.

8. Gram's Theorem

We are now driving at the heart of classical invariant theory; that is, why invariants were important to nineteenth century mathematicians. Our first observation is that once all the absolute invariants of $\Gamma \subset \mathbf{GL}(n; \Omega)$ are known, so are all the concomitants. For suppose that $\varphi : E_1 \times E_2 \times \cdots \times E_r \to T_q{}^p(E)$ is a simultaneous concomitant. From φ one can construct an absolute invariant as follows: for $u \in T_q{}^p(E)^* = T_p{}^q(E)$ and for $z_j \in E_j$, consider the form $\psi(z_1, z_2, \ldots, z_r, u) = \langle \varphi(z_1, z_2, \ldots, z_r), u \rangle$. If $\sigma \in \Gamma$

$$\begin{aligned}\psi(\sigma \cdot z_1, \ldots, \sigma \cdot z_r, \sigma \cdot u) &= \langle \varphi(\sigma \cdot z_1, \ldots, \sigma \cdot z_r), \sigma \cdot u \rangle \\ &= \langle \sigma^{-1} \cdot (\sigma \cdot \varphi(z_1 \ldots, z_r)), u \rangle \\ &= \psi(z_1, \ldots, z_r, u)\end{aligned}$$

so ψ is an absolute invariant. By letting $u_1, \ldots, u_k$ be a basis of $T_q{}^p(E)^*$, one has $\varphi(z_1, \ldots, z_r) = \Sigma \psi(z_1, \ldots, z_r, u_k)\, u_k$, and hence ψ uniquely determines φ.

Klein's definition of the geometry of $\Gamma \subset \mathbf{GL}(n; \Omega)$ is the totality of *algebraic* properties invariant under Γ. Such a property involves a certain number of mixed tensors $u, v, \ldots, w$ and is expressible by a finite number of algebraic relations between the components; e.g., a system of relations $K_h(u, v, \ldots, w) = 0$, where each K_h is a polynomial in the components of the tensors $u, v, \ldots, w$. In addition, the relations must have invariant significance; that is, for each $\sigma \in \Gamma$, $K_h(u, v, \ldots, w) = 0$ if and only if $K_h(\sigma \cdot u, \sigma \cdot v, \ldots, \sigma \cdot w) = 0$. The motivation for thc scarch for invariants is provided by Gram's Theorem, which says that a system of relations having invariant significance is equivalent to a system $K_h' = 0$ where the K_h' are *absolute invariants.* Before proving Gram's Theorem, we need a lemma.

Lemma. *Consider a tensor space $T_q{}^p(E)$, and let $e_1, \ldots, e_n$ be a basis of E. For any basis $y = (y_1, \ldots, y_n)$ of E, let $\sigma_y \in \mathbf{GL}(n; \Omega)$ be the transformation such that $\sigma_y(e_j) = y_j$. Then for $u \in T_q{}^p(E)$, $\sigma_y^{-1} \cdot u$ has components (on the corresponding basis of $T_q{}^p(E)$) which are absolute rational invariants of $u, y_1, \ldots, y_n$.*

Proof. By definition, if $\tau \in \Gamma$, $\sigma_{\tau.y} = \tau\sigma_y$. Hence $\sigma_{\tau.y}^{-1} = \sigma_y^{-1}\tau^{-1}$, so $\sigma_{\tau.y}^{-1} \cdot (\tau \cdot u) = \sigma_y^{-1} \cdot u$, and it follows that the components of $\sigma_y^{-1} \cdot u$ are absolute invariants.

Gram's Theorem. *Suppose the relations*

$$K_h(u_1, \dots, u_r) = 0 \qquad (1 \leqslant h \leqslant s, u_j \in T_{q_j}^{p_j}(E))$$

have invariant significance for $\mathbf{GL}(n; \Omega)$. *Then these relations are equivalent to a system of relations* $K_h'(u_1, \dots, u_r, y_1, \dots, y_n) = 0$, *where the* K_h' *are absolute invariants.*

Proof. By hypothesis, $K_h(u_1, \dots, u_r) = 0$ if and only if

$$K_h(\sigma \cdot u_1, \dots, \sigma \cdot u_r) = 0$$

for all $\sigma \in \mathbf{GL}(n; \Omega)$, so for all $y = (y_1, \dots, y_n)$ the relations $K_h(u_1, \dots, u_r) = 0$ imply $K_h(\sigma_y^{-1} \cdot u_1, \dots, \sigma_y^{-1} \cdot u_r) = 0$ which are of the form

$$K_h'(u_1, \dots, u_r, y_1, \dots, y_n) = 0.$$

But the K_h' are polynomials in the components of absolute invariants of $y_1, \dots, y_n, u_1, \dots, u_r$.

Conversely, if these relations hold for every choice of the y_j, then taking $y_j = e_j$ for all j gives back $K_h(u_1, \dots, u_r) = 0$.

9. Invariants of n-ary Forms: The Symbolic Method

An isomorphism between the space of homogeneous polynomials in n variables of degree r (n-ary forms of order r) and the space $\mathscr{S}^r(E)^*$ of symmetric multilinear forms of r vectors in an n-dimensional vector space E has been demonstrated in Chapter 1. Given any n-ary form of order $r, f(x) = \Sigma\beta_{r_1\cdots r_n}(\xi^1)^{r_1} \cdots (\xi^n)^{r_n}$ $(r_1 + r_2 + \cdots + r_n = r)$ in the components $(\xi^1, \dots, \xi^n)$ of x with respect to the basis $e_1, \dots, e_n$ of E, there is associated a unique multilinear symmetric form expressed as $\varphi(\overset{1}{x}, \dots, \overset{r}{x}) = \Sigma a_{\alpha_1\cdots\alpha_r}\overset{\alpha_1}{\underset{1}{\xi}}\overset{\alpha_2}{\underset{2}{\xi}} \cdots \overset{\alpha_r}{\underset{r}{\xi}}$ in the components $(\underset{j}{\xi}^1, \dots, \underset{j}{\xi}^n)$ of the $\underset{j}{x}$, the summation ranging over all multi-indices $\alpha = (\alpha_1, \dots, \alpha_r)$ with $1 \leqslant \alpha_j \leqslant n$ for $1 \leqslant j \leqslant r$; φ has the property that $\varphi(x, x, \dots, x) = f(x)$.

The coefficients a_α of φ are symmetric; that is, if $\pi \in \mathscr{G}_r$, $a_{\pi \cdot \alpha} = a_\alpha$. Let $\nu(\alpha) = (r_1, \dots, r_n)$ be the multi-index such that r_j is the number of terms in α equal to j. Fixing $(r_1, \dots, r_n)$ with $r_1 + r_2 + \cdots + r_n = r$, the number of α with $\nu(\alpha) = (r_1, \dots, r_n)$ is $r!/r_1!r_2! \cdots r_n!$. Hence the coefficient $\beta_{\nu(\alpha)}$ of the monomial $x^{\nu(\alpha)} = (\xi^1)^{r_1} \cdots (\xi^n)^{r_n}$ in f is precisely $(r!/r_1!r_2! \cdots r_n!)a_\alpha$. By virtue of the isomorphism of $\mathscr{S}^r(E)^*$ with $\mathscr{S}^r(E^*) = \underset{=}{s}(T_r^0(E))$, φ is uniquely associated with a symmetric covariant tensor of order r. We briefly describe this isomorphism. Let $\varphi \in (E^{\otimes r})^*$ satisfy $\varphi(\overset{1}{x} \otimes \cdots \otimes \overset{r}{x}) = \Sigma a_{\alpha_1 \cdots \alpha_r} \underset{1}{\xi}^{\alpha_1} \cdots \underset{r}{\xi}^{\alpha_r}$, and let $e_1{}^*, \ldots, e_n{}^*$ be the basis of E^* dual to $e_1, \ldots, e_n$. The image of φ is $\Sigma a_{\alpha_1 \cdots \alpha_r} e_{\alpha_1}^* \otimes \cdots \otimes e_{\alpha_r}^*$; in other words, the coefficients of φ determine the components of the covariant tensor which is the image of φ. This isomorphism carries $\mathscr{S}^r(E)^*$ onto $\mathscr{S}^r(E^*)$.

Given a certain number of generic n-ary forms $f_1, \ldots, f_k$ and a certain number of contravariant vectors, an invariant of these forms and vectors will, by definition, be a simultaneous invariant of these vectors and of the k symmetric covariant tensors corresponding to the forms f_i. The general problem is thus to find these invariants.

For definiteness, we shall restrict ourselves to two n-ary forms f and g of orders r and s and a single vector $x \in E$. Suppose

$$f = \sum (r!/r_1 ! \cdots r_n !)\, a_{r_1 \cdots r_n} (\xi^1)^{r_1} \cdots (\xi^n)^{r_n},$$

$$g = \sum (s!/s_1 ! \cdots s_n !)\, b_{s_1 \cdots s_n} (\xi^1)^{s_1} \cdots (\xi^n)^{s_n},$$

and $x = (\xi^1, \ldots, \xi^n)$, where $r_1 + \cdots + r_n = r$ and $s_1 + \cdots + s_n = s$. We seek all homogeneous polynomial invariants of f, g, and x of degree p in the $a_{\vec{r}}$ ($\vec{r} = (r_1, \ldots, r_n)$), degree q in the $b_{\vec{s}}$ ($\vec{s} = (s_1, \ldots, s_n)$), and degree m in the ξ^i. This is equivalent by polarization to finding all multilinear invariants φ of p covariant symmetric tensors $u_1, \ldots, u_p$ of order r, q covariant symmetric tensors $v_1, \ldots, v_q$ of order s, and m contravariant vectors $\overset{1}{x}, \ldots, \overset{m}{x}$. In turn we consider multilinear invariants ψ' of pr covariant vectors y_{ih}' ($1 \leqslant i \leqslant p$, $1 \leqslant h \leqslant r$), qs covariant vectors z_{jk}' ($1 \leqslant j \leqslant q$, $1 \leqslant k \leqslant s$), and m contravariant vectors $\overset{1}{x}, \ldots, \overset{m}{x}$, which are, for each i, symmetric in the y_{ih}' and, for each j, symmetric in the z_{jk}'. To find these, we apply the theorem on invariants of mixed tensors. In particular, a relation must exist between m, n, p, q, r, and s; namely, $gn = m - pr - qs$ for some integer g defined to be the weight of the invariant ψ'.

The restitution of ψ' to φ is carried out as follows. Let y_{ih}' have components $(\overset{i}{\eta}_{h1}, \overset{i}{\eta}_{h2}, \ldots, \overset{i}{\eta}_{hn})$ and replace, in each term of ψ', the monomial $(\overset{i}{\eta}_{1\alpha_1})(\overset{i}{\eta}_{2\alpha_2}) \cdots (\overset{i}{\eta}_{r\alpha_r})$ by $\underset{i}{a}_{r_1 \cdots r_n}$, where $(r_1, \ldots, r_n)$ is the multi-index $\nu(\alpha)$. Now do the same for the z_{jk}'. The $\underset{i}{a}_{r_1 \cdots r_n}$ determine the components of the symmetric tensor u_i of order $r = r_1 + r_2 + \cdots + r_n$. To restitute u_i to f, one simply suppresses the index i in the $\underset{i}{a}_{r_1 \cdots r_n}$.

The restitution may be simplified somewhat. In place of the distinct vectors y'_{ih}, one considers a single vector y_i' and does likewise for the z'_{jk}. If the components of y_i' are $(\overset{i}{\eta}_1, ..., \overset{i}{\eta}_n)$, one replaces in the expression for ψ' each monomial $(\overset{i}{\eta}_1)^{r_1}(\overset{i}{\eta}_2)^{r_2} \cdots (\overset{i}{\eta}_n)^{r_n}$ with respect to these components by $\underset{i}{a}_{r_1 \cdots r_n}$. Doing the same for the z'_{jk}, one obtains the following rule.

Aronhold's Rule. Every multilinear invariant φ of p n-ary forms of degree r, of q n-ary forms of degree s, and of m contravariant vectors is obtained by polarization from a polynomial invariant ψ of $p + q$ covariant vectors y_i', z_j' $(1 \leqslant i \leqslant p, 1 \leqslant j \leqslant q)$ and of m contravariant vectors $\overset{i}{x}$. ψ is of degree r with respect to each of the y_i', of degree s with respect to each of the z_j', and linear with respect to each of the $\overset{i}{x}$. φ is deduced from ψ by replacing each monomial $(\overset{i}{\eta}_1)^{r_1} \cdots (\overset{i}{\eta}_n)^{r_n}$ (resp. $(\overset{j}{\zeta}_1)^{s_1} \cdots (\overset{j}{\zeta}_n)^{s_n}$) by the component $\underset{i}{a}_{r_1 \cdots r_n}$ (resp. $\underset{j}{b}_{s_1 \cdots s_n}$) of the corresponding form.

The application of this rule is called the *symbolic method*, and ψ is called the *symbolic expression* of the invariant φ. One notes, for example, that the polynomial f, considered as a polynomial with respect to the $a_{r_1 \cdots r_n}$ and the components ξ^i of x is an absolute invariant (since $m - pr = 0$) whose symbolic expression is $\langle x, y' \rangle^r$. Finally we remark that Aronhold's rule can be generalized to include any number of symmetric covariant tensors and contravariant vectors.

Example 1. Let us find homogeneous invariants of weight $g = -2$ of a single n-ary quadratic form $f = \sum \alpha_{ij} \xi^i \xi^j$ $(\alpha_{ij} = \alpha_{ji})$. Since the relation $2p = -gn$ must hold between g and the degree p of the invariant, we take $p = n$. Hence we are looking for all multilinear invariants φ of n covariant symmetric tensors of order 2. By an earlier section we already know $\varphi(y'_{11}, y'_{12}, ..., y'_{n2})$ is a linear combination of products of two covariant brackets. It follows that for n vectors $y_1', ..., y_n' \in E^*$, the only non-trivial possibility for $\psi(y_1', ..., y_n')$ is $c[y_1' \cdots y_n']^2$ with $c \in \Omega$. Suppose y_i' has components $(\overset{i}{\eta}_1, ..., \overset{i}{\eta}_n)$; then $[y_1' \cdots y_n']^2$ is the determinant of the matrix whose (i, j)-th entry is $\overset{1}{\eta}_i \overset{1}{\eta}_j + \overset{2}{\eta}_i \overset{2}{\eta}_j + \cdots + \overset{n}{\eta}_i \overset{n}{\eta}_j$. But after two restitutions $\overset{k}{\eta}_i \overset{k}{\eta}_j$ becomes α_{ij}, so the only invariant in question is a scalar multiple of the discriminant $d = \det(\alpha_{ij})$ of the form f. The symbolic expression of the invariant d is $(n^n)^{-1}[y_1' \cdots y_n']$.

Example 2. Consider an n-ary quadratic form f and a linear form $l \in E^*$, and let us find all homogeneous invariants of degree $(n - 1)$ in the components of f and degree two in the components of l. Here $m = 0, p = n - 1, r = 2, q = 2$ and $s = 1$, so $g = -2$. The invariant obtained by the symbolic method has symbolic expression $[y_1' \cdots y_{n-1}' z']^2$. The relation $[y_1' \cdots y_n' z']^2 = 0$ has invariant significance; namely, that the hyperplane $l = 0$ is tangent to the quadric $f = 0$.

Example 3. The Jacobian of a system of n-ary forms. Given n-ary forms $f_1, \ldots, f_n$ of orders $r_1, \ldots, r_n$ and a vector $x \in E$, the Jacobian $J = \det(\partial_j f_i)$ has the following symbolic expression:

$$J = c[y_1' \cdots y_n']\langle y_1', x\rangle^{r_1-1} \cdots \langle y_n', x\rangle^{r_n-1}.$$

Note $\partial_j f_i$ is obtained by restitution from $r_i \overset{i}{\eta}_j \langle y_i', x\rangle^{r_i-1}$, so J is obtained by restitution of $(r_1 \cdots r_n) \det(\overset{i}{\eta}_j \langle x, y_i'\rangle^{r_i-1})$ which reduces to the former expression.

Example 4. The Hessian $H_f = \det(\partial_i \partial_j f)$ of an n-ary form f of order d has as its symbolic expression

$$c[y_1' \cdots y_n'][y_1'\langle x, y_1'\rangle^{d-2} \cdots y_n'\langle x, y_n'\rangle^{d-2}].$$

Example 5. As a final example, let us find a complete or fundamental system of invariants for two binary quadratic forms

$$f = \alpha_{11}(\xi^1)^2 + 2\alpha_{12}\xi^1\xi^2 + \alpha_{22}(\xi^2)^2, \qquad g = \beta_{11}(\xi^1)^2 + 2\beta_{12}\xi^1\xi^2 + \beta_{22}(\xi^2)^2,$$

and a contravariant vector $x = (\xi^1, \xi^2)$. That is, we mean to find a finite number of polynomial invariants $d_1, \ldots, d_r$ of f, g, and x such that every polynomial invariant of f, g, and x is a polynomial of $\Omega[d_1, \ldots, d_r]$. Consider an invariant φ, and suppose ψ is its symbolic expression. Applying the symbolic method, we may assume ψ is a monomial whose factors are among

$$[y_i' y_n'], [z_j' z_m'], [y_i' z_j'], \langle \overset{h}{x}, y_i'\rangle, \langle \overset{h}{x}, z_j'\rangle.$$

In each such product, every y_i' occurs twice, every z_j' occurs twice, and every $\overset{h}{x}$ occurs once. (It is needless to consider factors of the form $[\overset{h}{x}\overset{j}{x}]$ since these vanish after restitution.)

Assume first that ψ contains the factor $[y_1'y_2']$. Then it must contain a factor of one of the following forms:

(i) $[y_1'y_2']^2$
(ii) $[y_1'y_2'][y_1'a'][y_2'b']$
(iii) $[y_1'y_2'][y_1'a']\langle \overset{h}{x}, y_2'\rangle$
(iv) $[y_1'y_2']\langle \overset{h}{x}, y_1'\rangle\langle \overset{k}{x}, y_2'\rangle$.

By Example 1, (i) is the symbolic expression of $d_1 = \alpha_{11}\alpha_{22} - (\alpha_{12})^2$. Since $[y_1'y_2']\langle \overset{h}{x}, y_1'\rangle\langle \overset{k}{x}, y_2'\rangle$ and $[y_2'y_1']\langle \overset{h}{x}, y_2'\rangle\langle \overset{k}{x}, y_1'\rangle$ express the same invariant, the restitution of (iv) gives 0. Similarly, if $a' = b'$, then (ii) restitutes to 0. Hence suppose $a' \neq b'$. By Section 6,

$$[y_1'a'][y_2'b'] - [y_2'a'][y_1'b'] + [y_2'y_1'][a'b'] = 0$$

so that

$$\begin{aligned}\psi &= [y_1'y_2'][y_1'a'][y_2'b']\,\theta \\ &= [y_1'y_2'][y_1'b'][y_2'a']\,\theta - [y_1'y_2']^2\,[a'b']\,\theta\end{aligned}$$

from which it follows that 2φ has d_1 as a factor. The restitution of (iii) is similar, again giving that φ has d_1 as a factor. Thus if ψ has $[y_i'y_j']$ as a factor, then φ has d_1 as a factor.

By the same reasoning, if ψ has $[z_j'z_k']$ as a factor, then $d_2 = \beta_{11}\beta_{22} - (\beta_{12})^2$ is a factor of φ. Hence we may assume that neither $[y_i'y_j']$ nor $[z_j'z_k']$ occur as factors in ψ. Hence, if, for example, $[y_1'z_1']$ is a factor, then so is one of the following products:

(v) $[y_1'z_1']^2$
(vi) $[y_1'z_1'][y_i'z_1'][y_1'z_j']$
(vii) $[y_1'z_1'][y_i'z_1']\langle \overset{h}{x}, y_1'\rangle$
(viii) $[y_1'z_1']\langle \overset{h}{x}, y_1'\rangle\langle \overset{k}{x}, z_1'\rangle$.

Now $[y_1'z_1']^2$ is the symbolic expression of the invariant $d_{12} = \alpha_{11}\beta_{22} + \alpha_{22}\beta_{11} - 2\alpha_{12}\beta_{12}$ and $[y_1'z_1']\langle \overset{h}{x}, y_1'\rangle\langle \overset{k}{x}, z_1'\rangle$ that of the invariant $h = (\alpha_{11}\beta_{12} - \alpha_{12}\beta_{11})(\xi^1)^2 + (\alpha_{11}\beta_{22} - \alpha_{22}\beta_{11})\xi^1\xi^2 + (\alpha_{12}\beta_{22} - \alpha_{22}\beta_{12})(\xi^2)^2$. To restitute (vi) set

$$\begin{aligned}\psi &= [y_1'z_1'][y_i'z_1'][y_1'z_j']\,\theta \\ &= [y_1'y_i'][z_j'z_1'][y_1'z_1']\,\theta + [y_1'z_1']^2\,[y_i'z_j']\,\theta.\end{aligned}$$

Hence φ is, in this case, the sum of two invariants, the first having d_1 as a factor, the second having d_{12} as a factor. The restitution of (vii) gives a corresponding result. Hence, by induction on the number of factors of ψ, one can express φ as a sum of invariants, each having as a factor a monomial in d_1, d_2, d_{12}, and h, such that suppression of this monomial in each term yields an invariant whose symbolic expression contains no brackets. But an invariant of this kind must decompose into a product of factors of the form $\langle \overset{h}{x}, y_i' \rangle \langle \overset{k}{x}, y_i' \rangle$ or $\langle \overset{h}{x}, z_j' \rangle \langle \overset{k}{x}, z_j' \rangle$, which are the symbolic expressions of f and g, respectively. We summarize our results in the following theorem.

Theorem. *Every polynomial invariant of two binary quadratic forms f, g and a contravariant vector x is a polynomial in the invariants d_1, d_2, d_{12}, $h, f,$ and g.*

10. Invariants of Subgroups of $\mathbf{GL}(n; \Omega)$

Letting $\mathscr{G}_f$ act on $T_f^0(E) = E^{*\otimes f}$ in a manner analogous to its action on $T_0^f(E)$, one obtains a direct sum decomposition $T_f^0(E) = \Sigma_i e_i \cdot T_f^0(E)$, where each e_i is a projection onto a space of irreducible tensors and has the form $c_\alpha t$ for some Young's diagram Σ_α and some $t \in \Omega[\mathscr{G}_f]$. Recall that as an element of $T_0^f(E)^*$, $y_1' \otimes \cdots \otimes y_f'$ is characterized by

$$\langle x_1 \otimes \cdots \otimes x_f , y_1' \otimes \cdots \otimes y_f' \rangle = \langle x_1 , y_1' \rangle \cdots \langle x_f , y_f' \rangle,$$

so it follows that $\langle z, \pi \cdot z' \rangle = \langle \pi^{-1} \cdot z, z' \rangle$ if $\pi \in \mathscr{G}_f$. If $z' \in T_f^0(E)$ is a relative invariant for $\Gamma \subset \mathbf{GL}(n; \Omega)$, then, by the commutation property, so are $e_i \cdot z'$, $(b_\alpha t) \cdot z'$, and $t \cdot z'$, and, by the above relation and the definition of a_α,

$$\langle x_1 \otimes \cdots \otimes x_f , (a_\alpha b_\alpha t) \cdot z' \rangle = \langle a_\alpha \cdot (x_1 \otimes \cdots \otimes x_f), (b_\alpha t) \cdot z' \rangle.$$

Expressing $\langle z, z' \rangle$ in this way gives the *Young–Deruyts development* of z'. This development has several consequences with regard to the problem of deducing information about the invariants of subgroups from information about the invariants of $\mathbf{GL}(n; \Omega)$.

Theorem. *All invariants of $\Gamma \subset \mathbf{GL}(n; \Omega)$ of an arbitrary number of contravariant vectors $x \in E$ are known once one knows for Γ all the polynomial invariants of at most $(n - 1)$ contravariant vectors.*

Proof. With the above notations, let $u' = (b_\alpha t) \cdot z'$, where z' is a linear invariant of Γ on $T^0{}_f(E)$, and let us compute $\langle a_\alpha \cdot (x_1 \otimes \cdots \otimes x_f), u' \rangle$. Assuming that the k-th column of the frame α has β_k cases, consider, in place of $x_1 \otimes \cdots \otimes x_f$, the tensor $w = z_1 \otimes \cdots \otimes z_p$, with $z_i = y_1 \otimes \cdots \otimes y_{\beta_i}$ for $1 \leqslant i \leqslant p$, p being the number of columns of α. This defines a simultaneous polynomial invariant $v(y_1, \ldots, y_{\beta_1}) = \langle w, u' \rangle$ of β_1 contravariant vectors which uniquely determines

$$\langle a_\alpha \cdot (x_1 \otimes \cdots \otimes x_f), u' \rangle.$$

In fact, since $a_\alpha = (\Sigma p_1) \cdots (\Sigma p_{\beta_1})$, where p_i runs through all permutations of the i-th row of the diagram Σ_α corresponding to a_α,

$$\langle a_\alpha \cdot (x_1 \otimes \cdots \otimes x_f), u' \rangle$$

is just the multilinear invariant (in $x_1, \ldots, x_f$) obtained by polarizing v, first with respect to y_1, next with respect to y_2, and so forth, and hence is determined by v.

If $\beta_1 > n$ then $u' = 0$, so we may assume $\beta_1 = n$. Assume also that $t \cdot z' = x_1' \otimes \cdots \otimes x_f'$. Then $u' = b_\alpha \cdot (t \cdot z') = \underline{a}(x_1' \otimes \cdots \otimes x_n') \otimes w'$ with $w' \in T^0_{f-n}(E)$. Thus

$$\begin{aligned} \langle w, u' \rangle &= \langle y_1 \otimes \cdots \otimes y_n, \underline{a}(x_1' \otimes \cdots \otimes x_n') \rangle \langle z_2 \otimes \cdots \otimes z_p, w' \rangle \\ &= c[y_1 \cdots y_n] \langle z_2 \otimes \cdots \otimes z_p, w' \rangle. \end{aligned}$$

This relation can be seen directly or can be deduced from the first main theorem, since $\langle y_1 \otimes \cdots \otimes y_n, \underline{a}(x_1' \otimes \cdots \otimes x_n') \rangle$ is a multilinear invariant of $\mathbf{GL}(n; \Omega)$. Since $[y_1 \cdots y_n]$ is an invariant, $\langle z_2 \otimes \cdots \otimes z_p, w' \rangle$ must be an invariant of Γ of $y_1, \ldots, y_{\beta_2}$. Repeated application of this argument, therefore, allows one to consider only the case when $\beta_1 < n$, so the theorem is proved.

One obtains from the proof of this theorem another proof of the first main theorem for $\mathbf{GL}(n; \Omega)$, since every invariant I of at most $(n - 1)$ vectors is constant. In fact, if $e_1, \ldots, e_{n-j}$ ($j \geqslant 1$) are fixed independent vectors in E, and if $x_1, \ldots, x_{n-j}$ are also independent in E, there exists a $\sigma \in \mathbf{SL}(n; \Omega)$ (i.e., $\det \sigma = 1$) such that $\sigma(x_i) = e_i$. Thus $I(x_1, \ldots, x_{n-j}) = I(e_1, \ldots, e_{n-j})$ if $x_1, \ldots, x_{n-j}$ are independent. But if I is a polynomial, then, by the principle of irrelevancy, I must be constant.

As a second application, consider the extended group

$$\Gamma_\nu \subset \mathbf{GL}(n + \nu; \Omega)$$

whose elements are matrices of the form

$$\begin{pmatrix} \underline{A} & \underline{0} \\ \underline{B} & \underline{C} \end{pmatrix}$$

with $\underline{A} \in \Gamma \subset \mathbf{GL}(n; \Omega)$, $\underline{C} \in \mathbf{SL}(\nu; \Omega)$, and $\underline{B}$ arbitrary. For example, if $\Gamma = \mathbf{GL}(n; \Omega)$, then Γ_1 is called the affine group. If Γ is the orthogonal group (see the next example), Γ_1 is called the group of isometries of E.

Theorem. *To obtain the invariants of Γ_ν of any number of contravariant vectors in $\Omega^{n+\nu}$, add to the invariants of Γ the bracket $[x_1 \cdots x_{n+\nu}]$.*

Proof. By the last theorem it suffices to show that every polynomial invariant $I(x_1, ..., x_{n+\nu-j})$ of Γ_ν with $j \geqslant 1$ depends only on the coordinates of the x_i of index $\leqslant n$, and hence is uniquely determined by an invariant of Γ. Let $\mu = n + \nu$ and $x_i = (\xi_{1i}, ..., \xi_{\mu i})$ for $1 \leqslant i \leqslant \mu - j$. Suppose that the determinant

$$\Delta = \begin{vmatrix} \xi_{11} & \cdots & \xi_{1,\mu-j} \\ \vdots & & \vdots \\ \xi_{\mu-j,1} & \cdots & \xi_{\mu-j,\mu-j} \end{vmatrix}$$

is nonvanishing. Then it is possible to find some $\sigma \in \Gamma_\nu \cap \mathbf{SL}(n + \nu ; \Omega)$ of the form

$$\begin{pmatrix} 1 & & & 0 \\ & \cdot & & \\ & 0 & \cdot & \\ b_1 \cdots b_{\mu-j} & 0 \cdots 0 & & 1 \end{pmatrix}$$

such that each vector $y_i = \sigma \cdot x_i$ $(1 \leqslant i \leqslant \mu - j)$ has μ-th component 0. In other words, from the nonvanishing of Δ, $b_1, ..., b_{\mu-j}$ can certainly be chosen so that the relation $\eta_{\mu i} = b_1\xi_{1i} + \cdots + b_{\mu-j}\xi_{\mu-j,i} + \xi_{\mu i} = 0$ holds for the last component $\eta_{\mu i}$ of $y_i = (\eta_{1i}, ..., \eta_{\mu i})$. By the invariance, $I(y_1, ..., y_{\mu-j}) = I(x_1, ..., x_{\mu-j})$ so we get the polynomial relation $P(\xi_{hi}) = P(\eta_{hi})$ provided $\Delta \neq 0$, where $P(\xi_{hi}) = I(x_1, ..., x_{\mu-j})$. But by irrelevancy this proviso is unneeded, and one concludes that $P(\xi_{hi})$ never depends on $\xi_{\mu i}$; i.e., for arbitrary $x_1, ..., x_{n+\nu-j}$ of E, $I(x_1, ..., x_{n+\nu-j})$ depends only on the coordinates of each x_i of index $\leqslant n + \nu - 1$.

In general, the same argument can be carried out on the $(n + k)$-th row to show that $P(\xi_{hi})$ does not depend on $\xi_{n+k,i}$ for any i. One considers $\sigma \in \Gamma_\nu$ of the form

$$\begin{bmatrix} 1 & & & & 0 & & \\ \vdots & \ddots & & & & & \\ b_1 & \cdots\ b_{n+k-1} & 1 & b_{n+k+1} & \cdots\ b_{\mu-j} & 0 \cdots 0 \\ \vdots & & & & \ddots & & \\ 0 & & & & & & 1 \end{bmatrix}$$

and carries out the same argument.

Example. Suppose Ω is the complex number field $\mathbf{C}$ and $E = \mathbf{C}^n$ equipped with the bilinear form $(x \mid y) = \sum_i \xi_i \eta_i$ for $x, y \in E$. Let $\mathbf{O}(n)$ be the subgroup of $\mathbf{GL}(n)$ fixing this form; i.e., $\sigma \in \mathbf{O}(n)$ if and only if $(\sigma(x) \mid \sigma(y)) = (x \mid y)$. Since the relation ${}^t\sigma\sigma = \underline{I}$ must hold for each $\sigma \in \mathbf{O}(n)$, $\det \sigma = \pm 1$.

Theorem. *Any polynomial invariant of n vectors for $\mathbf{O}(n)$ is a polynomial in $[x_1 \cdots x_n]$ and the scalar products $(x_j \mid x_k)$. If the invariant has even weight, one may drop the bracket since $[x_1 \cdots x_n]^2 = \det(x_j \mid x_k)$.*

Proof. We use induction on n, the result being trivial for $n = 1$. From the Young–Deruyts development, it follows that a polynomial invariant in $x_1, \ldots, x_n$ is a combination of products of powers of $[x_1 \cdots x_n]$ and invariants of $n - 1$ of the vectors $x_1, \ldots, x_n$. One is therefore reduced to proving that any polynomial invariant f of $x_1, \ldots, x_{n-1}$ for $\mathbf{O}(n)$ is a polynomial in the scalar products $(x_j \mid x_k)$. There is always a $\sigma \in \mathbf{O}(n)$ which transforms $x_1, \ldots, x_{n-1}$ into vectors $x_1' = \sigma(x_1), \ldots, x_{n-1}' = \sigma(x_{n-1})$ which belong to the hyperplane H generated by the first $n - 1$ vectors of the natural basis of $E = \mathbf{C}^n$. We have therefore $f(x_1, \ldots, x_{n-1}) = f(x_1', \ldots, x_{n-1}')$. If f has odd weight, then $f = 0$, for the symmetry $\tau \in \mathbf{O}(n)$ with respect to H leaves all vectors in H invariant and has determinant -1, hence $f(x_1', \ldots, x_{n-1}') = -f(x_1', \ldots, x_{n-1}')$. On the other hand, any transformation $\sigma' \in \mathbf{O}(n - 1)$ can be extended to a transformation of $\mathbf{O}(n)$ leaving invariant the n-th basis vector, and therefore $f(x_1', \ldots, x_{n-1}') = f(\sigma'(x_1'), \ldots, \sigma'(x_{n-1}'))$. From the induction hypothesis it follows that if f is of even weight, $f(x_1', \ldots, x_{n-1}')$ is a polynomial in the $(x_j' \mid x_k') = (x_j \mid x_k)$, which ends the proof.

Chapter 3. Post-Hilbert Invariant Theory

1. The Finiteness Theorem

Actual computations of invariants of simple systems of forms and vectors had led to the conjecture that in every case all invariants were polynomials in a *finite* number of them. But this had only been proved (by Gordan, 1868) for binary forms. The proof by Hilbert (in 1890) of the general conjecture created therefore a big sensation, all the more so since Hilbert did not rely on any computational device, but on general finiteness arguments (the famous "Basissatz" which he proved in the same paper) and on a process associating an invariant form to a non-invariant one (the Ω-process of Cayley). We will give a proof, due to Nagata, of the finiteness theorem under more general hypotheses than those of Hilbert, including in particular invariants under all semi-simple groups over fields of characteristic zero, as well as finite groups. In 1900, Hilbert proposed as his fourteenth problem to generalize his result to any subgroup $\Gamma \subset \mathbf{GL}(n)$. This problem remained unsolved until 1958, when Nagata produced a counter example.

Let $R = K[a_1, \ldots, a_n]$ be a finitely generated algebra over an arbitrary field K. We are not assuming that the a_i's are algebraically independent over K. Γ will always be a group of algebra automorphisms of R satisfying the following assumption:

(i) the orbit under Γ of each $f \in R$ is contained in a finite dimensional subspace of R (over K).

Consequently, one may assume $\Gamma \subset \mathbf{GL}(r; K)$ for some r, for if $V = \Sigma_i(\Sigma_{s \in \Gamma} K(s \cdot a_i))$, then by (i) V is *always* a finite dimensional vector space stable by construction under Γ. The restriction $s \to s \mid V$ of Γ to V imbeds Γ in $\mathbf{GL}(V)$. For a basis $b_1, \ldots, b_r$ of V one certainly has that $K[b_1, \ldots, b_r] = R$, so we may always assume that the a_i give a basis for V.

The fundamental assumption of Nagata is the following:

(ii) If $s \to \underline{U}(s)$ is any finite dimensional linear representation of Γ which can be written

$$\underline{U}(s) = \begin{pmatrix} 1 & 0 & \cdots & 0 \\ c_2(s) & & & \\ \vdots & & \underline{V}(s) & \\ c_r(s) & & & \end{pmatrix}$$

then $\underline{U}$ is equivalent to

$$\underline{U}_1(s) = \begin{pmatrix} 1 & 0 & \cdots & 0 \\ 0 & & & \\ \vdots & & \underline{V}_1(s) & \\ 0 & & & \end{pmatrix}.$$

In other words, if $\underline{U}$ admits an invariant hyperplane, there exists an equivalent representation admitting an invariant hyperplane having a complementary line pointwise invariant under Γ. This is always the case if the representation $\underline{U}$ is completely reducible.

Let $I_\Gamma \subset R$ be the subalgebra of invariants of Γ. The Hilbert–Nagata theorem is the following.

Theorem. *The ring of invariants I_Γ with respect to a group Γ of algebra automorphisms of a finitely generated K-algebra R satisfying* (i) *and* (ii) *is finitely generated over K.*

The same proof gives a version for the class of groups studied in Chapter 2.

Corollary. *If $\Gamma \subset \mathbf{GL}(n; K)$ is a group of algebra automorphisms of $K[a_1, \dots, a_m]$ satisfying* (i), *if Γ acts rationally on $\Sigma_i K a_i$, and if every finite dimensional rational representation of Γ is completely reducible (or if Γ satisfies* (ii) *just for rational representations), then I_Γ is finitely generated over K.*

In Hilbert's original situation, K is the field of complex numbers, $\Gamma = \mathbf{SL}(n; K)$ acting by means of its tensor representations, and R is the K-algebra generated by the components of a finite number of mixed tensors $u_1, \dots, u_h$ over K^n. Thus we view R as the ring of polynomials of the space E spanned by $u_1, \dots, u_h$. If $P \in R$ and $\sigma \in \mathbf{SL}(n; K)$, $(\sigma \cdot P)$(components of u_j) $= P$(components of $\sigma^{-1} \cdot u_j$), since the only rational character on $\mathbf{SL}(n; K)$ is the constant character, so I_Γ is therefore the ring of polynomial (absolute) invariants of the action of Γ on E.

Examples of groups satisfying condition (ii) are:

(1) by Maschke's theorem, any finite group whose order is not divisible by the characteristic of K;

(2) for rational representations, $\mathbf{GL}(n; \Omega)$ and $\mathbf{SL}(n; \Omega)$;

(3) for rational representations, any semi-simple algebraic group $\Gamma \subset \mathbf{GL}(n; K)$ where K has characteristic 0;

(4) by a result of Weitzenböck, any 1-dimensional complex torus.

The proof of the finiteness theorem rests on the fundamental lemma of Nagata to be proven now. As always, assume Γ is a linear group satisfying (i) and (ii).

Lemma 1. *If $f \in R$, there exists an $f^* \in I_\Gamma \cap \Sigma_{s\in\Gamma} Ks \cdot f$ such that $f - f^* \in \Sigma_{s,t\in\Gamma} K(s \cdot f - t \cdot f)$.*

Proof. Let $M_f = \Sigma_{s\in\Gamma} Ks \cdot f$, and let $N_f = \Sigma_{s,t\in\Gamma} K(s \cdot f - t \cdot f)$. M_f and N_f are each finite dimensional Γ-stable subspaces of V. We thus prove the lemma by induction on $m = \dim_K M_f$. If $m = 0$, the lemma follows trivially, taking $f^* = 0$. Assume now that the lemma holds for $m < k$, and take any $f \in R$ for which $\dim_K M_f = k$. If $f \in N_f$ we may set $f^* = 0$. Otherwise, $f \notin N_f$, so $s \cdot f$ decomposes uniquely as $f + (s \cdot f - f)$; i.e., $M_f = N_f \oplus Kf$. Hence N_f is Γ-stable and is of codimension one, so by (ii) there exists a Γ-stable subspace P of M_f and an $f' \in I_\Gamma$ such that $M_f = Kf' \oplus P$.

Now if $f' \notin N_f$, then $M_f = Kf' \oplus N_f$, and thus $f = \lambda f' + g$ for some $g \in N_f$. In this case, we may set $f^* = \lambda f'$. When $f' \in N_f$, then $f = \lambda f' + h$ for some $h \in P$. But $M_h \subset P$ since P is stable, and it follows that $\dim_K M_h < k$. By the induction hypothesis, there exists an element $h^* \in I_\Gamma \cap M_h$ such that $h - h^* \in N_h$. But the invariance of f' implies that $s \cdot h - t \cdot h = s \cdot f - t \cdot f$, hence $N_h = N_f$. To complete the proof, take $f^* = h^*$. This gives $f - f^* = \lambda f' + (h - h^*) \in N_f$ and

$$f^* \in I_\Gamma \cap M_h \subset I_\Gamma \cap M_f$$

as desired.

Lemma 2. *Let $f_1, \ldots, f_r \in I_\Gamma$. Then $I_\Gamma \cap \sum_{i=1}^r Rf_i = \sum_{i=1}^r I_\Gamma f_i$.*

Proof. That $I_\Gamma \cap \sum_{i=1}^r Rf_i \supset \sum_{i=1}^r I_\Gamma f_i$ is immediate. To establish the opposite inclusion we shall use induction on r, the case $r = 0$ being trivial. Hence suppose the lemma holds if $r < k$, and let

$$f \in I_\Gamma \cap \sum_{i=1}^k Rf_i$$

where $f_1, \ldots, f_k \in I_\Gamma$. Write $f = \sum_{i=1}^k h_i f_i$, $h_i \in R$, and choose, by the fundamental lemma, an $h' \in N_{h_k}$ such that $h' + h_k \in I_\Gamma$. Since $f \in I_\Gamma$, $\sum_{i=1}^k (s \cdot h_i - t \cdot h_i) f_i = s \cdot f - t \cdot f = 0$ for all $s, t \in \Gamma$, hence $(s \cdot h_k - t \cdot h_k) f_k = -\sum_{i=1}^{k-1} (s \cdot h_i - t \cdot h_i) f_i$. It follows from this and the definition of N_{h_k} that $h'f_k = \sum_{i=1}^{k-1} h_i' f_i$, where each $h_i' \in R$, and hence that $f - (h_k + h')f_k = \sum_{i=1}^{k-1} (h_i - h_i')f_i$. Therefore

$$f - (h_k + h')f_k \in I_\Gamma \cap \sum_{i=1}^{k-1} Rf_i,$$

so we may apply the induction hypothesis to obtain, for each $i = 1,..., k-1$, an $h_i'' \in I_\Gamma$ such that $f - (h_k + h')f_k = \sum_{i=1}^{k-1} h_i'' f_i$. This completes the proof.

It is convenient now to break the proof up into two steps. We first consider the case when R is a graded ring and when Γ consists of graded automorphisms and deduce the general case from this. Note that here condition (i) is automatically fulfilled.

Definition. A K-algebra R is said to be *graded* if $R = \Sigma_{i\geqslant 0} R_i$ (direct) where:

(i) R_0 is a K-algebra;

(ii) each R_k is an R_0-module;

(iii) $R_i R_j \subset R_{i+j}$.

R is called *connected* if $R_0 = K$. The elements of R_i are called *homogeneous* of *degree i*. An automorphism σ of R is said to be *graded* if $\sigma(R_k) = R_k$ for each k.

Lemma 3. *Let* $R = \Sigma_{i\geqslant 0} R_i$ *be a graded K-algebra with* $R_0 = K$. *If the ideal* $R_+ = \Sigma_{i\geqslant 1} R_i$ *is finitely generated as an R-module, then R is finitely generated as a K-algebra.*

Proof. Write $R_+ = \Sigma_i R a_i$ and without loss of generality suppose $a_1,..., a_n$ are homogeneous of degree at least one. We will prove by induction that $R = R_0[a_1,..., a_n]$. Hence suppose that, for any homogeneous element $f \in R$ of degree $i \geqslant 0$, $f \in R_0[a_1,..., a_n]$. If $f \in R_{i+1}$ then $f = \Sigma_j g_j a_j$ where we may assume that each g_j is homogeneous of degree $\deg f - \deg a_j \leqslant i$. But then $g_j \in R_0[a_1,..., a_n]$ for $1 \leqslant j \leqslant n$ and hence so does f. Since $R_0 \subset R_0[a_1,..., a_n]$ the proof is complete.

Suppose now that R is a polynomial ring $K[T_1,..., T_m]$ in algebraically independent elements $T_1,..., T_m$ and Γ consists of graded automorphisms. Then I_Γ is graded, for if $f = f_1 + \cdots + f_k \in I_\Gamma$ and each $f_i \in R_i$, then the conditions $s \cdot f = f$ and $s \cdot f_i \in R_i$ for all $s \in \Gamma$ imply that each f_i is invariant. Let I_+ be the ideal in I_Γ consisting of all elements of positive degree. RI_+ is an ideal in R so by the "Hilbert Basissatz" there exist elements $f_1, \ldots, f_r$ of I_+ such that $RI_+ = \Sigma Rf_i$. Applying Lemma 2 we get that $RI_+ \cap I_\Gamma = \Sigma_i I_\Gamma f_i$, and from this that $I_+ = \Sigma I_\Gamma f_i$. We conclude that I_+ is finitely generated over I_Γ, and therefore that I_Γ is finitely generated over K. This finishes the proof in the graded case.

In the general case, let $R = K[a_1, \ldots, a_n]$ where $a_1, \ldots, a_n$ are linearly independent and let $R' = K[T_1, \ldots, T_n]$ be the ring of polynomials in the algebraically independent elements $T_1, \ldots, T_n$. Lift the action of Γ on R to R' in the following way: if, for $s \in \Gamma$, $s \cdot a_i = \Sigma_j s_{ij} a_j$, set $s \cdot T_i = \Sigma s_{ij} T_j$. Due to the fact that R' is a free algebra on $T_1, \ldots, T_n$, s can be extended to a graded automorphism of R'. Thus condition (i) is satisfied for this action of Γ on the graded ring R'. Therefore the ring I_Γ' of invariants of Γ in R' is finitely generated over K.

Let $\varphi : R' \to R$ be the K-algebra homomorphism defined by setting $\varphi(T_i) = a_i$ and $\varphi(\lambda) = \lambda$ if $\lambda \in K$. By construction, φ is *equivariant*; that is, $\varphi(s \cdot f) = s \cdot \varphi(f)$. Thus $\varphi(I_\Gamma')$ is a finitely generated subring of I_Γ. The theorem therefore follows from the next lemma.

Lemma 4. $\varphi(I_\Gamma') = I_\Gamma$.

Proof. Let $f = \varphi(f') \in I_\Gamma$. Since $f' \in R'$ there exists an $f^* \in I_\Gamma'$ such that $f' - f^* \in \Sigma_{s,t \in \Gamma} K(s \cdot f' - t \cdot f')$. The equivariance of φ and the invariance of f together imply that

$$\varphi(f' - f^*) \in \Sigma_{s,t\in\Gamma} K(\varphi(s \cdot f') - \varphi(t \cdot f')) = \Sigma_{s,t\in\Gamma} K(s \cdot \varphi(f') - t \cdot \varphi(f')) = \{0\}.$$

Therefore $f = \varphi(f^*) \in \varphi(I_\Gamma')$.

2. The Nagata Counterexample

Hilbert's fourteenth problem, to which we shall give a counterexample, may be stated as follows: *Let K be a field and Γ a subgroup of* $\mathbf{GL}(n; K)$. *If $x_1, \ldots, x_n$ are algebraically independent over K, then Γ acts as a group of (graded) automorphisms of $K[x_1, \ldots, x_n]$. Is the ring of invariants I_Γ of Γ in $K[x_1, \ldots, x_n]$ always finitely generated over K?* Now Γ acts uniquely as a group of automorphisms of the field of fractions $K(x_1, \ldots, x_n)$ relative to the *invariant subfield* L_Γ. Since

$$I_\Gamma = K[x_1, \ldots, x_n] \cap L_\Gamma$$

we may more generally ask: *If L is a subfield of $K(x_1, \ldots, x_n)$, is*

$$K[x_1, \ldots, x_n] \cap L$$

finitely generated over K? We show that the answer to this question is negative for some L, and we actually give a construction of a group Γ for which $L_\Gamma = L$.

Let $R = K[X_1,\ldots, X_r, Y_1,\ldots, Y_r]$ be the ring of polynomials in $2r$ indeterminates over an *infinite* field K, and let $E = K(X_1,\ldots, Y_r)$ denote the field of fractions of R. Set $t = Y_1 \cdots Y_r$, $u_i = Y_1 \cdots \hat{Y}_i \cdots Y_r$, and $v_i = u_i \cdot X_i$, and choose for $j = 1, 2, 3$ elements $a_{ji} \in K$ which are algebraically independent over the prime field of K. Finally, introduce for $j = 1, 2, 3$ algebraically independent elements (over K) $w_j = \Sigma_{i=1}^{r} a_{ji} v_i$. We shall partially prove the following theorem.

Theorem. *If* $r = n^2 \geqslant 16$ *and if* $L = K(w_1, w_2, w_3, t)$, *then* $I = R \cap L$ *is not finitely generated over* K.

Γ will be realizable as a subgroup of $\mathbf{GL}(2r; K)$ with $I_\Gamma = I$ and so I_Γ will not be finitely generated. The proof of this theorem will be based on a series of lemmas. Let $H = K[w_1, w_2, w_3]$.

Lemma 1. $I \subset \{\Sigma g_n t^{-n} : g_n \in H\}$.

Proof. First we establish that $I \subset B = K[t^{-1}][w_1, w_2, w_3, Y_1,\ldots, Y_r]$. Since $X_i = v_i Y_i t^{-1}$,

$$K[t^{-1}][X_1,\ldots, X_r, Y_1,\ldots, Y_r] = K[t^{-1}][v_1,\ldots, v_r, Y_1,\ldots, Y_r].$$

Also,

$$K[t^{-1}][v_1,\ldots, v_r, Y_1,\ldots, Y_r] = K[t^{-1}][w_1, w_2, w_3, v_4,\ldots, v_r, Y_1,\ldots, Y_r]$$

due to the fact that the a_{ji} are independent over the prime field. Hence $I \subset K(w_1, w_2, w_3, t) \cap K[t^{-1}][w_1, w_2, w_3, Y_1,\ldots, Y_r][v_4,\ldots, v_r]$; that is $I \subset$ (field of fractions of B) $\cap\, B[v_4,\ldots, v_r]$. But $v_4,\ldots, v_r$ are algebraically independent over B, so it follows easily that $I \subset B$.

Observe next that

$$K[t^{-1}][w_1, w_2, w_3, Y_1,\ldots, Y_r]$$
$$= K[w_1, w_2, w_3, t, t^{-1}][Y_2,\ldots, Y_r, Y_2^{-1},\ldots, Y_r^{-1}]$$

because $Y_1 = tY_2^{-1} \cdots Y_r^{-1}$ and $Y_i^{-1} = u_i t^{-1}$. Thus

$$I \subset K(w_1, w_2, w_3, t) \cap K[w_1, w_2, w_3, t, t^{-1}][Y_2,\ldots, Y_r, Y_2^{-1},\ldots, Y_r^{-1}],$$

which implies similarly that $I \subset K[w_1, w_2, w_3, t, t^{-1}]$ ($Y_2, \ldots, Y_r$ being algebraically independent over $K[w_1, w_2, w_3, t, t^{-1}]$). This completes the proof of Lemma 1.

For each $i = 1, \ldots, r$ let V_i denote the discrete valuation on E defined as follows: If $fg^{-1} \in E$ is written $fg^{-1} = Y_i^n pq^{-1}$ where $p, q \in R$ are polynomials relatively prime to Y_i and n is an integer, set $V_i(fg^{-1}) = n$. The set of elements in E for which $V_i \geqslant 0$ is called the *valuation ring* of V_i. Restricting V_i to L gives a valuation on L (denoted also by V_i) whose valuation ring we denote by $\mathcal{O}_i$. The ideal $\mathfrak{m}_i = \{f \in \mathcal{O}_i : V_i(f) \geqslant 1\}$ in $\mathcal{O}_i$ is maximal; indeed, $\mathcal{O}_i - \mathfrak{m}_i$ consists entirely of units.

Consider new elements $z_i = a_{3i}w_1 - a_{1i}w_3$ and $z_i' = a_{3i}w_2 - a_{2i}w_3$ of $H = K[w_1, w_2, w_3]$. Since $z_i = \Sigma_{j \neq i}(a_{3i}a_{1j} - a_{1i}a_{3j})\, v_j$ (with a corresponding expression valid for z_i'), it follows that

$$z_i = Y_i\, p_i(X_1, \ldots, \hat{Y}_i, \ldots, Y_r)$$

and $z_i' = Y_i\, p_i'(X_1, \ldots, \hat{Y}_i, \ldots, Y_r)$, where both p_i and p_i' are nonzero. Hence $V_i(z_i) = V_i(z_i') = 1$. One should also note that $V_i(w_3) = 0$, this because $w_3 = a_{3i}X_i u_i + Y_i\, p(X_1, \ldots, Y_r)$.

Sublemma. *w_3, $z_i t^{-1}$, $z_i' t^{-1}$ are algebraically independent (over K) modulo $\mathfrak{m}_i$. In other words, if $Q \in K[w_3, z_i t^{-1}, z_i' t^{-1}] \cap \mathfrak{m}_i$, then $Q = 0$.*

Proof. For if $Q \in K[w_3, z_i t^{-1}, z_i' t^{-1}] \cap \mathfrak{m}_i$, then setting $Y_i = 0$ gives $0 = Q(a_{3i}v_i, z_i t^{-1}, z_i' t^{-1})$. Multiplying by a sufficiently high power of t gives a relation

$$P(v_i, z_i, z_i', t) = t^s Q(a_{3i}v_i, z_i t^{-1}, z_i' t^{-1}) = 0.$$

But as v_i, z_i, z_i', and t are algebraically independent, $P = 0$, and thus $Q = 0$.

Let $\mathscr{P}_i$ be the ideal in H generated by z_i and z_i'.

Lemma 2. (a) *For each integer $m \geqslant 1$, $\mathscr{P}_i^m = \{f \in H : V_i(f) \geqslant m\}$.*
(b) *If $g_n \in H$, $V_i(\Sigma g_n t^{-n}) = \inf_n V_i(g_n t^{-n})$.*

Proof. To prove (a) we will use induction on m. We must first prove that $\mathscr{P}_i = \mathfrak{m}_i \cap H$. Suppose then $f \in H$ and $f \notin \mathscr{P}_i$. Since

$$K[z_i, z_i', w_3] = K[w_1, w_2, w_3],$$

one can write $f(w_1, w_2, w_3) = P(z_i, z_i', w_3) + h(w_3)$ where $V_i(P) \geqslant 1$

and $h(w_3) \neq 0$. By the sublemma $V_i(h(w_3)) = 0$, hence $V_i(f) = 0$, and therefore $f \notin \mathfrak{m}_i$.

Assuming now that (a) holds for m, we observe that to establish (a) for $m + 1$ it will suffice to prove that $\mathscr{P}_i^m - \mathscr{P}_i^{m+1} \subset \{f \in H : V_i(f) = m\}$. For $h \in \mathscr{P}_i^m - \mathscr{P}_i^{m+1}$ we have

$$h = \sum_{j=0}^{m} h_j z_i^j z_i'^{m-j} + Q$$

where $Q \in \mathscr{P}_i^{m+1}$ (thus $V_i(Q) > m$), $h_j \in K[w_3]$ for each j, and $h_j \neq 0$ for some j. Hence

$$ht^{-m} = \Sigma_j h_j (z_i t^{-1})^j (z_i' t^{-1})^{m-j} + Qt^{-m}.$$

Since $V_i(h) \geqslant m$, let us suppose that $V_i(ht^{-m}) > 0$. From $V_i(Qt^{-m}) > 0$, the inequality $V_i(\Sigma_j h_j (z_i t^{-1})^j (z_i' t^{-1})^{m-j}) > 0$ must obtain. But the sublemma says that this is impossible unless $h_1 = \cdots = h_m = 0$. This is a contradiction, so $V_i(h) = m$.

To prove (b), it suffices to show that, for $h_n \in H$, $V_i(\Sigma_{n \geqslant 0} h_n t^{-n}) = \inf_n V_i(h_n t^{-n})$. Let $d = \inf_n V_i(h_n t^{-n})$. By definition $V_i(h_n) = n + d$ for at least one n, say n_0, hence $n_0 + d \geqslant 0$. Now for all n, $V_i(h_n) \geqslant n + d$, so by (a), $h_n \in \mathscr{P}_i^{n+d}$ provided $n + d > 0$, and $h_n \notin \mathscr{P}_i^{n+d+1}$ for $n = n_0$ at least. Thus if $n + d \geqslant 0$ we may write

$$h_n = \sum_{j=0}^{n+d} f_{nj} z_i^j z_i'^{n+d-j} + Q_n$$

where each $f_{nj} \in K[w_3]$, $Q_n \in \mathscr{P}_i^{n+d+1}$, and for some j, $1 \leqslant j \leqslant n_0$, $f_{n_0 j} \neq 0$. If $n + d < 0$, define $Q_n = h_n$. Then

$$\Sigma_n h_n t^{-n-d} = \Sigma_{n \geqslant -d} \sum_{j=1}^{n+d} f_{nj} (z_i/t)^j (z_i'/t)^{n+d-j} + \Sigma_n Q_n t^{-n-d}.$$

Because

$$\Sigma_n Q_n t^{-n-d} = \Sigma_{n < -d} h_n t^{-n-d} + \Sigma_{n \geqslant -d} Q_n t^{-n-d}$$

one sees that $V_i(\Sigma_n Q_n t^{-n-d}) > 0$. But this implies, as in the proof of (a), that $V_i(\Sigma_n h_n t^{-n-d}) = 0$. Therefore $V_i(\Sigma h_n t^{-n}) = d$.

Lemma 3. *Let* $\mathfrak{A}_n = \bigcap_i \mathscr{P}_i^n$ $(n > 0)$. *Then*

$$I = \{\Sigma g_n t^{-n} : g_n \in \mathfrak{A}_n \text{ if } n > 0, \text{ and } g_n \in H \text{ if } n \leqslant 0\}.$$

Proof. We have already seen that if $f \in I$, $f = \Sigma g_n t^{-n}$ where each $g_n \in H$. It is sufficient, by Lemma 2(a), to show that $V_i(g_n) \geqslant n$ if $n > 0$. But this follows from the fact that if $f \in I$, $V_i(f) \geqslant 0$, and hence $V_i(g_n t^{-n}) \geqslant 0$ for each n by Lemma 2(b). The converse follows from the fact that I is contained in the ring

$$K[X_1, \ldots, X_r, Y_1, \ldots, Y_r, Y_1^{-1}, \ldots, Y_r^{-1}],$$

and in that ring the set of elements f such that $V_i(f) \geqslant 0$ for each i is R.

Lemma 4. *Let γ be a curve in the projective plane S of K having $r = n^2 \geqslant 16$ multiple points of order at least m in generic position. Then:*

(i) *the degree of γ is greater than $m\sqrt{r}$, and*

(ii) *given $r' > \sqrt{r}$, there exists a curve γ' satisfying the conditions above for γ for which $r' > \deg \gamma' \cdot m^{-1} > \sqrt{r}$.*

The proof of Lemma 4 will not be included here. For the proof, the reader is referred to Nagata [*4*] or Séminaire Bourbaki, Vol. 1958–59, n° 175.

The points $P_i = (a_{1i}, a_{2i}, a_{3i})$ of S defined for $i = 1, \ldots, r$ are by assumption in generic position. Since w_1, w_2, w_3 are algebraically independent over K, we may regard $H = K[w_1, w_2, w_3]$ as the homogeneous coordinate ring of S. $\mathscr{P}_i$ is the prime ideal in H corresponding to the point P_i.

We can now prove that I is not finitely generated. In fact, suppose $I = K[f_1, \ldots, f_m]$. It is convenient to assume, and we may by Lemma 3, that every $f_j = h_j t^{-j}$ where $h_j \in \mathfrak{A}_j$ and is homogeneous in w_1, w_2, w_3. Set $r_j = (\deg h_j) j^{-1}$, $r^* = \inf r_j$, and note that $r^* > \sqrt{r}$ by (i) of Lemma 4 and the definition of $\mathfrak{A}_j$. Since for any monomial

$$f = f_1^{i_1} \cdots f_m^{i_m} = h_1^{i_1} \cdots h_m^{i_m} t^{-i_1 - 2i_2 - \cdots - mi_m}$$

the relation $\deg f \geqslant r^*(i_1 + 2i_2 + \cdots + mi_m)$ holds, and as $f \in \mathfrak{A}_n$ with $n = i_1 + 2i_2 + \cdots + mi_m$, we have for any n and any homogeneous $a_n \in \mathfrak{A}_n$, $(\deg a_n) n^{-1} \geqslant r^*$ since the f_i generate I. But by Lemma 4(ii) there exists $a_n \in \mathfrak{A}_n$ such that $(\deg a_n) n^{-1} < r^*$. This is a contradiction, and therefore I is not finitely generated over K.

To complete the description of the counterexample, we will define a group Γ of automorphisms of R, $\Gamma \subset \mathbf{GL}(2r; K)$, such that $I_\Gamma = I$.

Indeed, let Γ be the group given by all matrices of the form

$$\sigma = \begin{pmatrix} \underline{B}_1 & & 0 \\ & \ddots & \\ 0 & & \underline{B}_r \end{pmatrix}$$

where

$$\underline{B}_i = \begin{pmatrix} c_i & c_i b_i \\ 0 & c_i \end{pmatrix}$$

subject to the conditions that $\sum_{i=1}^{r} b_i a_{ji} = 0$ ($j = 1, 2, 3$) and $c_1 \cdots c_r = 1$. Γ acts uniquely as a group of graded automorphisms of R such that $\sigma \cdot X_i = c_i(X_i + b_i Y_i)$ and $\sigma \cdot Y_i = c_i Y_i$ for all $\sigma \in \Gamma$. This action extends uniquely to the quotient field E of R making Γ into a group of automorphisms of E. Let $L_\Gamma \subset E$ be the field of invariants. We will show that $L_\Gamma = L$, and from this it follows that $I_\Gamma = R \cap L$.

By definition w_1, w_2, w_3, t are all invariant; hence $L \subset L_\Gamma$. Let $\Gamma_1 \subset \Gamma$ be the subgroup defined by setting $b_i = 0$ if $i \geqslant 5$ and $c_j = 1$ for all j. By assumption Γ_1 is infinite. Now we have

$$E = K(w_1, w_2, w_3, t, X_4, \ldots, X_r, Y_1, \ldots, Y_{r-1}) \supset L_{\Gamma_1}$$
$$\supset K(w_1, w_2, w_3, t, X_5, \ldots, Y_{r-1}).$$

By the next lemma and the fact that Γ_1 is infinite,

$$L_{\Gamma_1} = K(w_1, w_2, w_3, t, X_5, \ldots, Y_{r-1}).$$

Lemma. Let $K(x)$ *be a simple transcendental extension of* K, *and let* K' *be an intermediate field*; i.e., $K \subset K' \subset K(x)$. *If* H *is an infinite group of automorphisms of* $K(x)$ *leaving the elements of* K' *invariant, then* $K' = K$.

Proof. Suppose $K' \neq K$. By Lüroth's theorem ("Modern Algebra," Vol. I by Van der Waerden), K' is a simple transcendental extension of K, say $K' = K(\mathcal{O})$, and $K(x)$ is a finite algebraic extension of K'. This implies that the Galois group of $K(x)$ relative to $K(\mathcal{O})$ must be finite and thus contradicts the fact that H is infinite.

Now let $\Gamma_2 \supset \Gamma_1$ be the subgroup of Γ defined by $b_i = 0$ if $i \geqslant 6$ and $c_i = 1$ for all i. $L_{\Gamma_1} \supset L_{\Gamma_2} \supset K(w_1, w_2, w_3, t, X_6, \ldots, Y_{r-1})$, hence, by the last argument, $L_{\Gamma_2} = K(w_1, w_2, w_3, X_6, \ldots, Y_{r-1})$. Continue in this manner, getting that $L_{\Gamma'} = K(w_1, w_2, w_3, t, Y_1, \ldots, Y_{r-1})$ for the subgroup Γ' given by $c_i = 1$ for all i. Thus $L_\Gamma \subset K(w_1, w_2, w_3, t, Y_1, \ldots, Y_{r-1})$. But Γ acts on $K(w_1, w_2, w_3, t)[Y_1, \ldots, Y_{r-1}]$ with $\sigma \cdot Y_i = c_i Y_i$, and the invariant subfield of this action is just the base field $K(w_1, w_2, w_3, t)$, which proves that $L_\Gamma \subset L$.

Chapter 4. Introduction to the Hilbert–Mumford Theory

1. The Affine Case

The central problem in the recent work of Mumford on "geometric invariant theory" can be formulated in very vague terms as follows. Given an algebraic variety V and an algebraic group Γ acting algebraically on V, we may consider the set V/Γ of *orbits* of the points of V under Γ; is it possible, in a "natural" way, to endow that set with a structure of algebraic variety? One immediately realizes that even before attempting to give answers to that question, one first must face the task of giving general definitions for the words used in its formulation, and this immediately leads us into the deep waters of modern algebraic geometry. Needless to say, we will not embark on such an ambitious undertaking and will be content with merely skimming the surface by showing, under simplified assumptions, how the problem is closely linked with invariant theory, and giving a rough idea of how Mumford has been led to his powerful methods by a skillful development of a very original idea first introduced by Hilbert in his second big paper on invariant theory in 1893. For further information, we must refer the reader to Mumford's beautiful and difficult book.

We start with what is called the "affine case"; it is not very interesting in itself but has the merit of involving no trouble with definitions and of immediately introducing the connection with invariant theory. We consider an algebraically closed field K (if one wants to keep close to geometric intuition, one may take K to be the complex field), and an affine algebraic variety in the most naive sense, namely the set V of points $x = (x_1, \ldots, x_n)$ in some affine space K^n which satisfy a family of polynomial equations $P_\alpha(x) = 0$, where the P_α belong to the ring $A = K[X_1, \ldots, X_n]$. The group Γ will be here a subgroup of $\mathbf{GL}(n; K)$ which leaves V globally invariant.

Now the fundamental idea of affine algebraic geometry is to associate to V the ring R of "regular functions" on V, i.e., the restrictions to V of the polynomials on K^n; this is clearly isomorphic to the ring $A/\mathfrak{a}$, where $\mathfrak{a}$ is the ideal of all polynomials P vanishing on V (beware that $\mathfrak{a}$ contains the P_α but is not in general the ideal $\mathfrak{b}$ generated by them; $\mathfrak{a}$ is only the "root" of $\mathfrak{b}$, i.e., consists of polynomials some power of which is in $\mathfrak{b}$). The fundamental fact is that conversely R entirely determines V (Hilbert's Nullstellensatz): Namely, to every point $a \in V$ is associated the maximal ideal $\mathfrak{m}_a$ of R consisting of all regular functions f

such that $f(a) = 0$ (or equivalently the K-algebra homomorphism $g \to g(a)$ of R into K having $\mathfrak{m}_a$ as its kernel), and *every* K-homomorphism of R into K has that form, i.e., one has, in a "functorial" fashion, $V = \text{Hom}_K(R, K)$ (note that by definition a K-homomorphism $g : R \to K$ is such that $g(1) = 1$, hence is *surjective* since $K \subset R$). It is thus equivalent to say that Γ leaves invariant V, or leaves invariant the ring R under the natural action of Γ on R (Chapter 1); but this is again the same as saying that Γ, acting naturally on A, leaves the *ideal* $\mathfrak{a}$ globally invariant.

Functions on the set V/Γ with values in K are in one to one correspondence with functions defined on V and *invariant* by Γ. Any sensible definition of V/Γ as an algebraic variety will demand that "regular" functions on V/Γ lift back to "regular" functions on V; this means that the ring of "regular functions" on V/Γ must be the ring I_Γ of (absolute) *invariants* under Γ in R. This immediately imposes a restriction on Γ: namely, rings of "regular functions" such as R are always finitely generated over K, hence I_Γ must have that property, and we know from the Nagata counterexample that this is not always the case. We shall therefore assume that Γ satisfies conditions (i) and (ii) of Nagata described in Chapter 3, which ensures that I_Γ is finitely generated over K. We may then associate to the ring I_Γ the algebraic variety $W = \text{Hom}_K(I_\Gamma, K)$ as seen above. In addition, there is a natural mapping $\pi : V \to W$ which corresponds to the injection $j : I_\Gamma \to R$, namely the mapping $g \to g \circ j$ of $\text{Hom}_K(R, K)$ into $\text{Hom}_K(I_\Gamma, K)$; in terms of "points" (or maximal ideals), that mapping sends the maximal ideal $\mathfrak{m}_a$ on its intersection $\mathfrak{m}_a \cap I_\Gamma$ with I_Γ. We have to investigate if there is a "natural" bijection $h : V/\Gamma \to W$ such that the diagram

$$\begin{array}{ccc} & V & \\ {\scriptstyle p}\swarrow & & \searrow{\scriptstyle \pi} \\ V/\Gamma & \xrightarrow[h]{} & W \end{array}$$

is commutative, p being the natural mapping. The definition of h is obvious enough; every element of I_Γ being invariant under Γ, the intersections of $\mathfrak{m}_a$ and of all $\mathfrak{m}_{\sigma \cdot a}$ with I_Γ, where $\sigma \in \Gamma$, are the same. Hence π can be factored through V/Γ. What remains to be seen is if h is bijective.

"exceptional" vectors x were first systematically studied by n his 1893 paper, under the name of "Nullforms" (Hilbert was terested in spaces X of covariant symmetric tensors). Vectors here at least one invariant homogeneous F is such that $F(x) \neq 0$ d "*semi-stable points*" by Mumford. They can be equivalently ized by the following lemma, which uses the language of the opology in X:

a 2. *Suppose Γ satisfies the conditions* (i) *and* (ii) *of Nagata in* 3. *In order that $x \neq 0$ be semi-stable, it is necessary and sufficient* (*Zariski*) *closure of the orbit $O(x)$ in X should not contain the*

d, if F is a nonconstant homogeneous invariant polynomial such) $\neq 0$, then F is equal to a constant $\neq 0$ on $O(x)$ by definition, so equal to that constant on the Zariski closure $\overline{O(x)}$; however, $= 0$, 0 does not belong to $\overline{O(x)}$. Conversely, suppose $0 \notin \overline{O(x)}$ ly Lemma 1 to the Zariski-closed invariant subsets in X, 0} and $W_2 = \overline{O(x)}$; there is an invariant polynomial f such that and $f(z) = 1$ on W_2. But we may write $f = F_0 + F_1 + \cdots + F_m$, $_j$ is the homogeneous component of degree j of f, and we know h F_j is invariant under Γ (Chapter 1). The condition $f(0) = 0$ that the constant $F_0 = 0$; the other F_j take the value 0 at 0, and one of them must be $\neq 0$ on $\overline{O(x)}$, hence at x, which proves s semi-stable.

E HILBERT–MUMFORD CRITERION

possibility of using Lemma 2 to obtain an *explicit* criterion erizing "Nullforms" (or semi-stable points) derives from an idea ert. He was considering only the case in which $K = \mathbf{C}$, and $\mathbf{L}(n; \mathbf{C})$, hence Γ is a Lie group. To express that the point 0 lies losure (for the usual topology) of the orbit $O(x)$ of a point $x \neq 0$ Γ, he showed that it was necessary and sufficient that 0 should lie in the closure of the orbit of x under some *one-parameter* $p\ t \to \gamma(t)$ of $\mathbf{SL}(n)$; as $\gamma(t) \cdot x$ belongs to $X - \{0\}$ for any finite f $t \in \mathbf{C}$, the only possibility was that $\gamma(t) \cdot x$ *tends to* 0 *as t tends to* *t at infinity*. It must be admitted that the details of this argument ng "branches" of algebraic functions of t and the way in which e permuted around "branch points" are rather messy and alto-

In the first place, π (hence h) is always *surjective*; in other words, given any maximal ideal $\mathfrak{n}$ in I, there is a maximal ideal $\mathfrak{m}$ in R such that $\mathfrak{m} \cap I_\Gamma = \mathfrak{n}$. Indeed, by Hilbert's Basissatz, $\mathfrak{n}$ is generated by a finite number of invariants $f_1, \ldots, f_r$, and by Lemma 2 of Chapter 3, the ideal $\mathfrak{b} = R\mathfrak{n}$ in R generated by the f_j is such that $\mathfrak{b} \cap I_\Gamma = \mathfrak{n}$. Taking a maximal ideal $\mathfrak{m}$ of R containing $\mathfrak{b}$, we have that $\mathfrak{m} \cap I_\Gamma \supset \mathfrak{n}$, and as $\mathfrak{n}$ is maximal and $\mathfrak{m}$ does not contain 1, we have $\mathfrak{m} \cap I_\Gamma = \mathfrak{n}$.

To say that h is *injective* means that for any maximal ideal $\mathfrak{n}$ in I, all the points $a \in V$ such that $\mathfrak{m}_a \cap I_\Gamma = \mathfrak{n}$ belong to the *same* orbit; with the same notation, this means that all the maximal ideals containing $\mathfrak{b} = R\mathfrak{n}$ correspond to points of the same orbit; in other words, that orbit is the *algebraic subvariety* of V defined by the ideal $\mathfrak{b}$. We thus find a second necessary condition, namely that *all* orbits under Γ in V be algebraic subvarieties; another way to express this is to say that they are *closed* sets in the *Zariski topology* of V (where the closed sets are precisely the algebraic subvarieties of V; when $K = \mathbf{C}$, this topology is coarser than the usual one).

We now can show that this condition is also *sufficient* for h to be injective (Chevalley–Iwahori–Nagata). Indeed, we have to prove that if $\mathfrak{b}_1$, $\mathfrak{b}_2$ are the ideals of R corresponding to two different orbits, it is impossible that $\mathfrak{b}_1 \cap I_\Gamma = \mathfrak{b}_2 \cap I_\Gamma$. It is clear that $\mathfrak{b}_1$, $\mathfrak{b}_2$ are invariant under Γ. The result follows from the more general

Lemma 1. *Let $\mathfrak{b}_1$, $\mathfrak{b}_2$ be two invariant ideals in R corresponding to two Zariski-closed invariant sets W_1, W_2 without common points. Then there exists a function $f \in I_\Gamma$ such that $f(x) = 0$ in W_1 and $f(x) = 1$ in W_2.*

The fact that $W_1 \cap W_2 = \varnothing$ means there is no maximal ideal containing both $\mathfrak{b}_1$ and $\mathfrak{b}_2$, hence $\mathfrak{b}_1 + \mathfrak{b}_2 = R$, i.e., there exist $g_1 \in \mathfrak{b}_1$, $g_2 \in \mathfrak{b}_2$ such that $g_1 + g_2 = 1$. Hence, for s, t in Γ, $s \cdot g_1 - t \cdot g_1 = -(s \cdot g_2 - t \cdot g_2)$, and therefore

$$\Sigma_{s,t \in \Gamma} K(s \cdot g_1 - t \cdot g_1) \subset \mathfrak{b}_2 \cap (\Sigma_{s \in \Gamma} Ks \cdot g_1) \subset \mathfrak{b}_2 \cap \mathfrak{b}_1 .$$

By Lemma 1 of Chapter 3, there is a function $f \in I_\Gamma \cap \Sigma_{s \in \Gamma} Ks \cdot g_1$ such that $g_1 - f \in \Sigma_{s,t \in \Gamma} K(s \cdot g_1 - t \cdot g_1) \subset \mathfrak{b}_2$. Hence, for $x \in W_1$, $g_1(x) = s \cdot g_1(x) = 0$ for all $s \in \Gamma$, and $f(x) = 0$; on the other hand, for $y \in W_2$, $g_1(y) - f(y) = 0$. But $g_1(y) + g_2(y) = 1$ and $g_2(y) = 0$, hence $f(y) = 1$.

We shall see below in Section 3 examples where the orbits under Γ may not be Zariski-closed.

2. The Projective Case

When we replace affine varieties by projective varieties in the general problem considered in Section 1, we meet problems of geometric significance, which had already by the 19th century attracted the interest of mathematicians. An example is the "classification" of plane projective curves of a given order r, meaning that all curves which are deduced from one another by projective transformations should be put in the same class. Now if X is the vector space consisting of 0 and all ternary forms f of degree r, two such forms which differ by a constant factor define the same curve, hence the set which we are considering is the projective space $\mathbf{P}(X)$, and we look for the *orbits* in that space under the action of the group $\Gamma = \mathbf{PSL}(3; K)$ of projective transformations in the plane. As soon as $r \geqslant 3$, these orbits depend on "parameters," and to make this idea precise, we are again led to see if it is possible to put a "reasonable" algebraic variety structure on the set $\mathbf{P}(X)/\Gamma$; the answer, as we shall see, is "no"; some "special" curves of order r have to be discarded before the remaining set of orbits can be made into an algebraic variety. Such elementary "classification" problems point the way towards far more difficult ones such as the problem of "moduli" of curves solved in Mumford's book.

The only projective varieties which we shall consider henceforth are the projective spaces $\mathbf{P}(X)$, X being a *vector space* of dimension N over K; Γ will be a subgroup of $\mathbf{SL}(N; K)$ (we should take it as a subgroup of $\mathbf{PSL}(N; K)$, but homothetic mappings leave $\mathbf{P}(X)$ pointwise invariant, so the actions are the same); it will be subject to additional restrictions later on. We would like to repeat our procedure in the affine case by considering "regular functions" on $\mathbf{P}(X)$ invariant under Γ; the trouble is that there is no such thing as a "regular function," other than constants. Indeed, a function on $\mathbf{P}(X)$ can be considered as a function on $X - \{0\}$, which is constant on each ray $\{\xi a;\ a \neq 0,\ \xi \in K^*\}$; but no nonconstant polynomial on X has that property. In other words, it is meaningful (and important) to consider a projective "hypersurface" in $\mathbf{P}(X)$, consisting of all points which are images of points $x \in X - \{0\}$ which satisfy an equation $f(x) = 0$, where f is a homogeneous polynomial, but that hypersurface is not the set of points where a "regular function" vanishes. In order to get functions at all, we must be content with functions which are only defined on Zariski-open dense subsets of $\mathbf{P}(X)$; for instance, if f has degree m, and $\xi^1, \ldots, \xi^N$ are the coordinates of $x \in X - \{0\}$, we may consider the function $f(x)/(\xi^j)^m$, which is homo-

geneous of degree 0, hence constant on e
open subset U_j of $X - \{0\}$ defined by ξ
sets which together cover $X - \{0\}$, and o
the intersection $U_j \cap U_k$ the two functio
which is defined at each point as well as i

If we want to look for a "variety of o
then be to cover $X - \{0\}$ by open subsets
mappings, and *stable* under Γ, i.e., contai
points. If V_j' is the image of V_j in $\mathbf{P}(X)$,
to consider the "subvariety of orbits" V_j'
varieties (which is the case for the imag
above, which are isomorphic to K^{N-1}) we
restrictions on Γ, apply to each V_j' the me
finally $\mathbf{P}(X)/\Gamma$, we would have to "glue to
V_j'/Γ, something which in fact is rather eas
enough "abstract" algebraic geometry. Sin
such knowledge, we skip this "gluing proces
described in Mumford's book). To obtain
invariant under homothetic mappings an
method is to consider a nonconstant *homoge*
(under Γ), and take for V the set where $F($
such a V has an image V' in $\mathbf{P}(X)$ which is
degree of F, the ring of "regular functions
of the rational functions P_{mr}/F^m, where m ru
and, for each m, P_{mr} runs through all ho
degree mr (that ring is isomorphic to the r
nomials on X to the hypersurface defined by

3. Nullforms and Semi-Stable Points

At this point we stumble on the diffic
proceed with our program we should cove
invariant sets V' defined by the various ho
nomials F. But this would only be possible
$x \neq 0$ in X for which $F(x) = 0$ for *every*
polynomial invariant F. However, simple
vectors exist in general: For instance, if X
forms over K^n, and $\Gamma = \mathbf{SL}(n; K)$, the only
invariants on X are powers of the discriminan
Example 1); hence *all* homogeneous noncons
vanish at the *degenerate* quadratic forms (i.e.,

gether not very satisfactory. Following Mumford, we shall see how Hilbert's arguments can be made entirely rigorous and purely algebraic.

The field K being, as always, algebraically closed and having characteristic 0, we observe that $\Gamma = \mathbf{SL}(n; K)$ is an algebraic group; as an affine subvariety of K^{n^2}, it is defined by the principal ideal generated by $\det(Y_{ij}) - 1$ in the ring of polynomials $K[Y_{11}, \ldots, Y_{nn}]$ in n^2 indeterminates, and it is well known that this ideal is prime, hence the ring $A = K[Y_{11}, \ldots, Y_{nn}]/(\det(Y_{ij}) - 1)$ of "regular functions" on Γ is an integral domain. We suppose X has dimension N, and therefore has a ring of regular functions which is the ring of polynomials $R = K[X_1, \ldots, X_N]$. Now let $x \in X$ be a vector $\neq 0$; to the (rational) mapping $s \to s \cdot x$ of Γ into X corresponds a K-algebra homomorphism $\varphi : R \to A$: to each polynomial $u \in R$, it associates the "regular function" $\varphi(u)$ on Γ which, at the "generic" point (y_{ij}) (where y_{ij} is the natural image of Y_{ij} in A) takes the value $u((y_{ij}) \cdot x)$; and conversely the mapping $s \to s \cdot x$ is exactly the mapping $\mathrm{Hom}(\varphi, 1_K) : \mathrm{Hom}_K(A, K) \to \mathrm{Hom}_K(R, K)$. The kernel $\mathfrak{a}_x$ of φ is a prime ideal of R, since A is an integral domain; it defines the subvariety $\overline{O(x)}$ (Zariski closure) in X; the ring of regular functions on that variety is the integral domain $R_x = R/\mathfrak{a}_x$, and we may thus identify R_x to a subalgebra of A. Suppose now that the point 0 belongs to $\overline{O(x)}$; this means that it corresponds to a maximal ideal $\mathfrak{m}$ in R_x *which is not the intersection of* R_x *and a maximal ideal of* A.

The key lemma of commutative algebra on which Hilbert's proof now relies is the following one:

Lemma 3. *Let K be a field of characteristic* 0, *B a finitely generated K-algebra which is an integral domain, E its field of fractions, F a field which is a finitely generated extension of E. Let $\mathfrak{m}$ be a maximal ideal in B; then there exists a complete discrete valuation ring V containing B and whose field of fractions L contains F, such that if $\mathfrak{n}$ is the maximal ideal of V, $\mathfrak{n} \cap B = \mathfrak{m}$.*

Suppose first $F = E$. As B is noetherian, $\mathfrak{m}$ is finitely generated; let $x_1, \ldots, x_n$ generate the ideal $\mathfrak{m}$; we may assume that $x_1 \neq 0$. Let C be the subring of E generated by $x_2/x_1, \ldots, x_n/x_1$; then $x_j \in x_1 C$ for $j > 1$, and therefore $\mathfrak{m}C = x_1 C$ is a principal ideal in C. It follows from Krull's Hauptidealsatz that the prime ideals $\mathfrak{p}$ of C which are minimal among those containing $x_1 C$ are of height 1, i.e., no prime ideal $\neq 0$ may be properly contained in $\mathfrak{p}$. On the other hand, the intersection $\mathfrak{p}C_{\mathfrak{p}} \cap B$ contains $\mathfrak{m}$ and does not contain the unit element, hence is equal to $\mathfrak{m}$. Let now S be the integral closure of C in K; a classical theorem of E. Noether proves that S is a finitely generated C-module, hence is noetherian. Furthermore the ring of fractions $S_{\mathfrak{p}}$ is the integral closure of $C_{\mathfrak{p}}$; from the choice of $\mathfrak{p}$, $C_{\mathfrak{p}}$ has Krull dimension 1, hence the same is true of $S_{\mathfrak{p}}$ which is finite over $C_{\mathfrak{p}}$; if $\mathfrak{r}$ is a maximal ideal of $S_{\mathfrak{p}}$ containing $\mathfrak{p}S_{\mathfrak{p}}$, the local ring $(S_{\mathfrak{p}})_{\mathfrak{r}}$ is integrally closed, noetherian, and has dimension 1, which means it is a discrete valuation ring V whose maximal ideal $\mathfrak{n}$ is such that $\mathfrak{n} \cap C_{\mathfrak{p}} = \mathfrak{p}C_{\mathfrak{p}}$. Hence $\mathfrak{n} \cap B = \mathfrak{m}$.

If F is a finitely generated extension of E, we replace V by the valuation ring V' of a valuation on F extending the valuation on E corresponding to V; as F is finitely generated, it is well-known that V' is still a discrete valuation ring. Finally, we replace V' by its completion, and we are done.

We now apply Lemma 3 to $B = R_x$ (E being the field of fractions of R_x, F the field of fractions of A), which is finitely generated over E. We thus obtain a complete discrete valuation ring V containing R_x, whose maximal ideal intersects R_x in $\mathfrak{m}$, and whose field of fractions contains F; but the assumption on $\mathfrak{m}$ means that V *does not contain* A.

We next observe that a complete discrete valuation ring which is an algebra over a field K of characteristic 0 necessarily is the *ring of formal power series in one indeterminate* $L[[T]]$, where L is a field containing K; its maximal ideal consists of power series without constant term; and its field of fractions is the field $L((T))$ of power series in T with a finite number of terms having negative exponents.

We can summarize what we have done by saying that we have a commutative diagram of homomorphisms of K-algebras

$$\begin{array}{ccc} A & \xrightarrow{\theta} & L((T)) \\ {\scriptstyle\varphi}\uparrow & & \uparrow \\ R & \xrightarrow[\psi]{} & L[[T]] \end{array}$$

where the second vertical arrow is the natural injection, and by assumption $\theta(A) \not\subset L[[T]]$. The mapping $s \to s \cdot x$ of Γ into X can by assumption be written $\underline{Z} \to \underline{U}(\underline{Z}) \cdot x$, where the elements of the $N \times N$ matrix $\underline{U}(\underline{Z})$ are rational functions (with coefficients in K) of the elements of the $n \times n$ matrix $\underline{Z}$. Let $\underline{W}(T)$ be the $N \times N$ matrix over $L((T))$ whose elements are those of $\underline{U}((\theta(y_{ij})))$. Then the preceding definitions and constructions show that we have the two following properties:

(i) *the elements of* $\underline{W}(T)$ *do not all belong to* $L[[T]]$;

(ii) *the* N *coordinates of the vector* $\underline{W}(T) \cdot x$ (which *a priori* belong to $L((T)))$ *are series in* $L[[T]]$ *without constant term.*

We now use the fact that $L[[T]]$ is a principal ideal ring, hence by the theory of elementary divisors applied to the $n \times n$ matrix $\underline{Y}(T) = (\theta(y_{ij}))$ with elements in $L((T))$, there exist two $n \times n$ unimodular matrices $\underline{P}$, $\underline{Q}$ over $L[[T]]$, which have inverses with elements in $L[[T]]$, and are such that $\underline{Y}(T) = \underline{P}\underline{D}\underline{Q}$, where $\underline{D} = \underline{D}(T)$ is a diagonal matrix

$$\underline{D}(T) = \begin{pmatrix} T^{\alpha_1} & 0 & \cdots & 0 \\ 0 & T^{\alpha_2} & \cdots & 0 \\ \cdot & \cdot & \cdot & \cdot \\ 0 & 0 & \cdots & T^{\alpha_n} \end{pmatrix} \tag{1}$$

the α_j being positive or negative integers such that $\alpha_1 + \alpha_2 + \cdots + \alpha_n = 0$.

We have $\underline{U}(\underline{Y}(T)) = \underline{U}(\underline{P})\,\underline{U}(\underline{D})\,\underline{U}(\underline{Q})$, $\underline{U}$ being a homomorphism of $\mathbf{SL}(n; K)$ into $\mathbf{SL}(N; K)$; furthermore, for a second indeterminate T', we have $\underline{D}(TT') = \underline{D}(T)\,\underline{D}(T')$, hence $\underline{U}(\underline{D}(TT')) = \underline{U}(\underline{D}(T))\underline{U}(\underline{D}(T'))$. If m is a sufficiently large integer and $\underline{G}(T) = T^m\underline{U}(\underline{D}(T))$, the matrix $\underline{G}(T)$ has polynomial elements and also satisfies $\underline{G}(TT') = \underline{G}(T)\,\underline{G}(T')$. Therefore, the argument used in the proof of the theorem of Chapter 2, Section 1, shows that there exists an invertible $N \times N$ matrix $\underline{A}$ with elements in K, such that the matrix $\underline{A}\,\underline{U}(\underline{D})\,\underline{A}^{-1}$ has diagonal form

$$\underline{D}'(T) = \begin{pmatrix} T^{\beta_1} & 0 & \cdots & 0 \\ 0 & T^{\beta_2} & \cdots & 0 \\ \cdot & \cdot & \cdot & \cdot \\ 0 & 0 & \cdots & T^{\beta_N} \end{pmatrix} \tag{2}$$

the β_k being again positive or negative integers such that $\beta_1 + \beta_2 + \cdots + \beta_N = 0$. The result (i) means that the β_k are *not all* 0.

We can write $\underline{U}(\underline{P})\,\underline{A}^{-1} = \underline{R}_0 + T\underline{R}_1$, $\underline{A}\,\underline{U}(\underline{Q}) = \underline{S}_0 + T\underline{S}_1$, where $\underline{R}_0$ and $\underline{S}_0$ are invertible matrices with elements in L, and $\underline{R}_1$ and $\underline{S}_1$ are matrices with elements in $L[[T]]$. We have

$$\underline{D}'(T) \cdot (\underline{A}\,\underline{U}(\underline{Q}) \cdot x) = \underline{A}\,\underline{U}(\underline{P})^{-1} \cdot (\underline{W}(T) \cdot x) \tag{3}$$

and if we put $y = \underline{S}_0 \cdot x$ (which is a vector in the space $X_{(L)} = X \otimes_K L$), it is clear that the k-th component of the vector on the left-hand side of (3) has the form $T^{\beta_k}(y_k + Tz_k)$ where $z_k \in L[[T]]$. As by property (ii) above, all coordinates of the vector on the right-hand side of (3) belong to $TL[[T]]$, we see that we must have

$$\beta_k \geqslant 1 \textit{ for all indices } k \textit{ such that } y_k \neq 0. \tag{4}$$

Conversely, if there is an extension L of K, a K-homomorphism $\theta : A \to L((T))$ such that, for $\underline{W}(T)$ defined as above, the β_k are not all 0 and condition (4) holds, the composite mapping $\theta \circ \varphi$ sends all the indeterminates X_j into $TL[[T]]$ whereas $\theta(A) \not\subset L[[T]]$. The intersection of R_x with the maximal ideal of $L[[T]]$ is therefore the maximal ideal corresponding to the point 0, hence 0 belongs to the closure of the orbit of x.

When K is the complex field, it may be shown that formal power series may be replaced by convergent ones (in a neighborhood of 0), and $t \to \underline{Y}(1/t)$ is then the one-parameter Lie subgroup of $\mathbf{SL}(n; \mathbf{C})$ which Hilbert had in mind.

In general, Mumford shows that one may replace L by K in the preceding argument; furthermore, relying on a result of Iwahori, he proves that the same criterion for "nullforms" holds when Γ is replaced by any reductive algebraic group over K, and has generalized it when the projective space $\mathbf{P}(X)$ is replaced by any proper (over K) algebraic scheme. In addition, he has proved that if a point x is such that not only $\beta_k \leqslant 0$ when $y_k \neq 0$ for all possible "one parameter subgroups" of Γ (which by (4) means x is *semi-stable*), but even $\beta_k < 0$ under the same conditions, then x is what he calls a "properly stable" point; this means that (with the notations of Section 2) the orbits of Γ in the affine open subvariety V' of $\mathbf{P}(X)$ are Zariski-*closed* and the stabilizer of any point of V' has dimension 0; in particular the results of Section 1 may be applied to V', and (by the "gluing process" we have alluded to) one may finally obtain a "variety of orbits" when one only considers the action of Γ on the open subset of all "properly stable" points.

5. Examples

Let us (following Hilbert) determine completely the "nullforms" in the simplest of the cases which were at the center of interest of classical invariant theory, X being the space of *n-ary forms* of a given degree r (or equivalently, symmetric covariant tensors of order r), with $n = 2$ or $n = 3$.

(A) *Binary forms*

Here $N = r + 1$, the "coordinates" of a binary form

$$x = a_0\xi_1{}^r + a_1\xi_1^{r-1}\xi_2 + \cdots + a_r\xi_2{}^r$$

being a_0, a_1,..., a_r; with the preceding notations, if

$$\underline{D}(T) = \begin{pmatrix} T^{-\alpha} & 0 \\ 0 & T^{\alpha} \end{pmatrix}$$

where $\alpha > 0$ the matrix $\underline{U}(\underline{D}(T))$ transforms a_k into $a_k' = a_k T^{(2k-r)\alpha}$ for $0 \leqslant k \leqslant r$. The condition (4) can therefore only be satisfied if $a_k = 0$ for $2k - r \leqslant 0$, and the "nullforms" are all transforms (under $\mathbf{SL}(2; K)$) of such forms; if binary forms of order r are identified (up to a factor) with systems of r points (counted with multiplicities) in the projective line, the nullforms are those systems of points containing a *point of multiplicity* $\geqslant 1 + (r/2)$ *if r is even, of multiplicity* $\geqslant (r + 1)/2$ *if r is odd.*

(B) *Ternary forms*

Here $N = \frac{1}{2}(r + 1)(r + 2)$, and a ternary form

$$x = \Sigma a_{\nu_1\nu_2\nu_3}\xi_1^{\nu_1}\xi_2^{\nu_2}\xi_3^{\nu_3} \qquad (\text{with } \nu_1 + \nu_2 + \nu_3 = r)$$

has as "coordinates" the coefficients $a_{\nu_1\nu_2\nu_3}$ for all systems (ν_1, ν_2, ν_3) with $\nu_1 + \nu_2 + \nu_3 = r$. If $\underline{D}(T) = \text{diag}(T^{\alpha_1}, T^{\alpha_2}, T^{\alpha_3})$ with $\alpha_1 + \alpha_2 + \alpha_3 = 0$, $a_{\nu_1\nu_2\nu_3}$ is multiplied by $T^{\alpha_1\nu_1+\alpha_2\nu_2+\alpha_3\nu_3}$, and condition (4) is satisfied if and only if $a_{\nu_1\nu_2\nu_3} = 0$ for all triples such that

$$\alpha_1\nu_1 + \alpha_2\nu_2 + \alpha_3\nu_3 \leqslant 0. \tag{5}$$

Hilbert gives a graphical representation of this condition in the following way: He considers in a plane an equilateral triangle ABC each side of which has length r, the lines carrying AB, BC and CA being oriented in the respective senses from A to B, B to C and C to A. A point M in the plane is associated with the system of coordinates $(\zeta_1, \zeta_2, \zeta_3)$ where $\zeta_i = \overline{M_i M}$, M_1, M_2, M_3 being the points where the parallels to CA, AB, BC through M meet respectively, BC, CA, and AB. To the triple (ν_1, ν_2, ν_3) of integers is associated in the plane the point having these numbers as triangular coordinates. Observe that lines passing through the center 0 of the triangle have as equation $u_1\zeta_1 + u_2\zeta_2 + u_3\zeta_3 = 0$, with $u_1 + u_2 + u_3 = 0$, and that if the line contains one of the points with integral coordinates, the u_j are rational numbers, hence may be taken as integers. We then see that we obtain nullforms by considering an arbitrary line D through 0 and all triples (ν_1, ν_2, ν_3)

corresponding to points *on one side of* D (or on D), and taking in the form x all $a_{\nu_1\nu_2\nu_3} = 0$ corresponding to these triples. All other forms transformed from the latter by an element of $\mathbf{SL}(3; K)$ are nullforms as well, and all nullforms are obtained in that way. Furthermore, one may only consider lines D which contain *no* point (ν_1, ν_2, ν_3) with integral coordinates such that $\nu_1 + \nu_2 + \nu_3 = r$, with the possible exception of 0: for if we consider such a line D_0, and a line D sufficiently close to it, the points such that $\nu_1 + \nu_2 + \nu_3 = r$ and which are on one side of D are the *same* as those on one side of D; when one takes D_0 instead of D, this amounts to annihilating some additional coefficients, and therefore the nullforms thus obtained are *particular cases* of the ones corresponding to D. For each D one has to consider both sides of D; furthermore, by rotations of $2\pi/3$ or $4\pi/3$ leaving invariant ABC, one gets equivalent nullforms, so that the types of nonequivalent forms is smaller than one would expect.

We carry out the Hilbert construction for $r = 2, 3, 4, 5$ and give the corresponding types of curves corresponding to the nullforms.

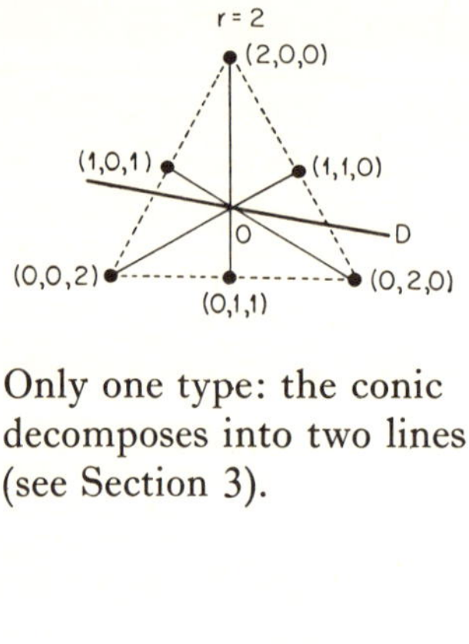

Only one type: the conic decomposes into two lines (see Section 3).

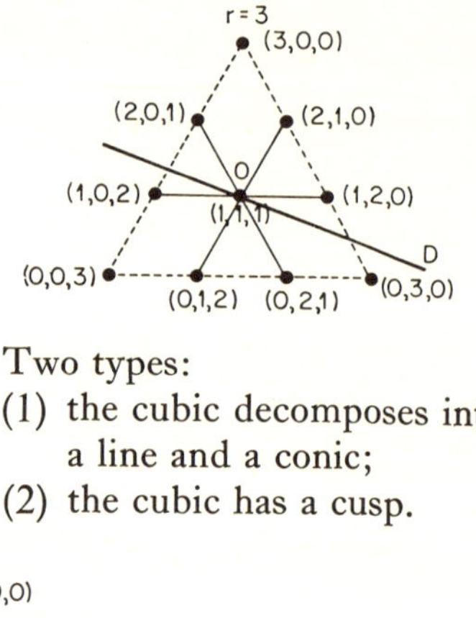

Two types:
(1) the cubic decomposes into a line and a conic;
(2) the cubic has a cusp.

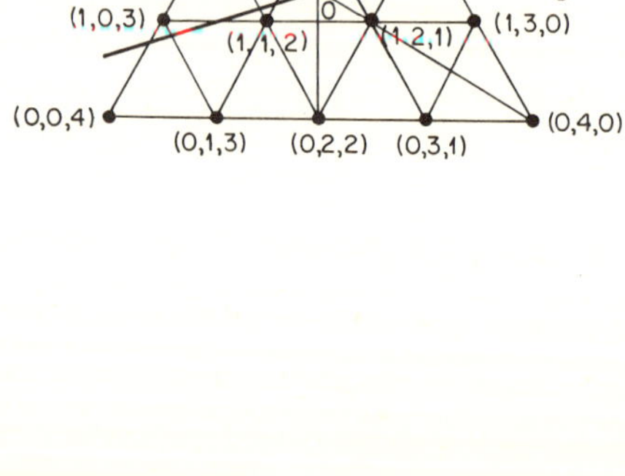

Two possible lines D. Three types:

(1) the quartic decomposes into a double line and a conic;

(2) the quartic decomposes into a cubic and an inflexional tangent to that cubic;

(3) the quartic has a triple point.

(The first case may be considered as a degeneracy of the second.)

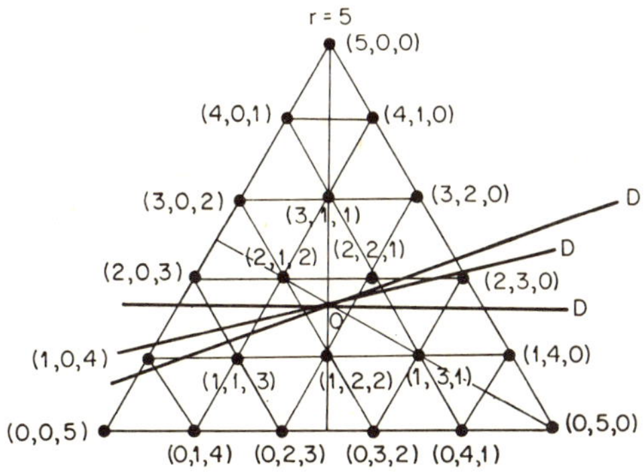

Three possible lines D. Four types:

(1) the quintic decomposes into a double line and a cubic;

(2) the quintic has a quadruple point;

(3) the quintic has a triple point with only one tangent;

(4) the quintic decomposes into a quartic with a double point and a line tangent to the quartic at the double point.

(The first case may be considered as a degeneracy of the fourth.)

6. Applications of Nullforms to Problems of Explicit Determination of Invariants

Hilbert's proof of the finiteness theorem (Chapter 3, Section 1) was, (as is Nagata's modification of that proof which we have presented), a purely *existential* one, and did not provide any means of determining, in a finite number of steps (bounded by an *a priori* computable number), a fundamental system of homogeneous polynomial invariants generating the whole ring of invariants (for an example of such a system, see Chapter 2, Section 9, Example 5). Hilbert wrote his 1893 paper in part to answer that criticism; using the concept of nullform and his "Nullstellensatz" (which he proved in that same paper), he showed how one could, at least theoretically, give a method which in every case was bound to lead to an explicit fundamental system of invariants. We shall now briefly describe that method.

The assumptions being the same as in Section 4, we first observe that it is the same thing to say that a homogeneous polynomial in $R = K[X_1, \ldots, X_N]$ is an absolute invariant under $\mathbf{SL}(n; K)$ or a relative invariant under $\mathbf{GL}(n; K)$; this second interpretation will be more convenient. It will be enough to show that the ideal RI_+ of R generated by nonconstant homogeneous invariants has a system of generators of degrees *bounded by a computable number* M; for then the symbolic method can, in a *finite* number of steps (which can be computed in terms of n, N, and the action of $\mathbf{SL}(n; K)$ on X), determine all such invariants.

Hilbert reduces that problem to the following one:

(P) Show that there is a computable number M' such that for every semi-stable element $x_0 \in X$ (i.e., not a nullform), there is a homogeneous polynomial invariant P (depending on x_0) of degree $\leqslant M'$ such that $P(x_0) \neq 0$.

Indeed, suppose that problem is solved, and consider, in R, the ideal $\mathfrak{J}$ generated by all polynomials P (corresponding to all semi-stable vectors $x_0 \in X$); then the algebraic variety in K^N defined by that ideal consists exactly of the nullforms. As by definition, all invariants vanish for nullforms, the Nullstellensatz shows that some power of every invariant is in $\mathfrak{J}$. As we may (at least theoretically) pick a finite system of generators of $\mathfrak{J}$ of degree $\leqslant M'$, we are then reduced to forming a system of generators of the "root" of $\mathfrak{J}$; Hilbert's own proof of his Nullstellensatz shows how this (again theoretically) may be done explicitly.

To solve problem (P), Hilbert first proves the following lemma:

Lemma 4. *Let $x_0 \neq 0$ be any vector in X, and let $B \subset K[Y_{11}, \ldots, Y_{nn}]$ be the subring generated by the N coordinates of the vector $(Y_{ij}) \cdot x_0$ (which are polynomials in the Y_{ij} of fixed degree p). In order that x_0 be semi-stable, a necessary and sufficient condition is that the element* $\det(Y_{ij})$ *in $K[Y_{11}, \ldots, Y_{nn}]$ be integral over the subring B.*

To prove necessity, suppose x_0 is semi-stable. Then there exists at least one nonconstant homogeneous polynomial invariant P such that $P(x_0) \neq 0$. If $g > 0$ is the weight of P, we have the identity

$$P((Y_{ij}) \cdot x_0) = \det(Y_{ij})^g \, P(x_0)$$

and as $(P(x_0))^{-1} \in K$, this is an equation of integral dependence for $\det(Y_{ij})$ over B.

Conversely, suppose we have an equation of integral dependence over B for $\det(Y_{ij})$:

$$(\det(Y_{ij}))^m + P_1((Y_{ij}) \cdot x_0)(\det(Y_{ij}))^{m-1} + \cdots + P_m((Y_{ij}) \cdot x_0) = 0 \qquad (6)$$

where the P_j are polynomials in $X_1, \ldots, X_N$, which obviously may be taken homogeneous and nonconstant since $\det(Y_{ij})$ is homogeneous. Use contradiction, supposing that x_0 is a nullform: then, by the Hilbert–Mumford criterion of Section 4, it would be possible to substitute for the indeterminate matrix (Y_{ij}) in (6) a matrix $\underline{D}(T)$ of type (1) and get for the rational functions $P_j(\underline{D}(T) \cdot x_0)$ *polynomials without constant term in* T; as $\det(\underline{D}(T)) = 1$, this is obviously impossible.

To apply this lemma, observe that as B is finitely generated, there exist, among the N coordinates of $(Y_{ij}) \cdot x_0$, a number of *algebraically independent* elements $z_1, \ldots, z_r$ (over K) such that the ring B is integral over its subring $B' = K[z_1, \ldots, z_r]$ (E. Noether's normalization lemma); hence if x_0 is semi-stable, it follows from Lemma 4 that $\det(Y_{ij})$ is also *integral over* B'. As B' is an integrally closed ring, the minimal equation of $\det(Y_{ij})$ over the *field of fractions* $E' = K(z_1, \ldots, z_r)$ of B' has its coefficients in B'. We are going to derive from this that in that equation

$$(\det(Y_{ij}))^q + Q_1((Y_{ij}) \cdot x_0)(\det(Y_{ij}))^{q-1} + \cdots + Q_q((Y_{ij}) \cdot x_0) = 0 \qquad (7)$$

each $Q_k(x_0)$ is the value at x_0 of an *invariant* polynomial. Indeed, for any element $s \in \mathbf{SL}(n; K)$, we may substitute in (7) the matrix $(Y_{ij})s$ for the matrix (Y_{ij}) and still have 0 on the left-hand side; however, as $\det((Y_{ij})\, s) = \det(Y_{ij})$, this is only possible if, for every k,

$$Q_k((Y_{ij}) \cdot (s \cdot x_0)) = Q_k((Y_{ij}) \cdot x_0),$$

and replacing (Y_{ij}) by the unit matrix, we get

$$Q_k(s \cdot x_0) = Q_k(x_0) \qquad \text{for every } s \in \mathbf{SL}(n; K). \qquad (8)$$

However, from Nagata's Lemma 1 in Chapter 3, we get that there exists an invariant homogeneous polynomial Q_k^* in $K[X_1, \ldots, X_n]$ such that $Q_k - Q_k^*$ is a linear combination, with coefficients in K, of polynomials $s \cdot Q_k - t \cdot Q_k$ (for s, t in $\mathbf{SL}(n; K)$). This implies, by (8), that $Q_k(x_0) = Q_k^*(x_0)$.

Finally, replacing (Y_{ij}) by the unit matrix in (7), we get

$$1 + Q_1^*(x_0) + \cdots + Q_q^*(x_0) = 0$$

and therefore, at least one of the $Q_k{}^*(x_0)$ must be $\neq 0$. But $Q_k{}^*$ is an invariant polynomial of weight k (by homogeneity), hence of degree kn/p (and of course it must be 0 if kn/p is not an integer); as $kn/p \leqslant qn/p$, we see that problem (P) will be solved provided we solve the following problem:

(P′) Show that there is a computable number M'' such that the degree q of Eq. (7) is bounded by M'' for *every* semi-stable element $x_0 \in X$.

This, however, is not difficult. It relies on the following elementary lemma:

Lemma 5. *Let $H_1, \ldots, H_{h+1}$ be homogeneous polynomials of degree m in h indeterminates $u_1, \ldots, u_h$ over K. Then, if $L \geqslant h(m+1)^{h-1}$, there is an identical relation*

$$\Sigma C_{\alpha_1\alpha_2\cdots\alpha_{h+1}} H_1^{\alpha_1} \cdots H_{h+1}^{\alpha_{h+1}} = 0 \tag{9}$$

between the H_i's, with $\alpha_1 + \alpha_2 + \cdots + \alpha_{h+1} = L$ *and coefficients $C_{\alpha_1\alpha_2\ldots\alpha_{h+1}}$ in K not all* 0.

Indeed, suppose that lemma is proved. Complete the sequence $z_1, \ldots, z_r$ with $n^2 - r$ elements $z_{r+1}, \ldots, z_{n^2}$, taken from among the Y_{ij}^p, such that $(z_1, \ldots, z_{n^2})$ is a transcendence basis of the field $K(Y_{11}, \ldots, Y_{nn})$ over K. As $z_{r+1}, \ldots, z_{n^2}$ are algebraically independent over $K(z_1, \ldots, z_r)$, the minimal polynomial of $\det(Y_{ij})$ over $K(z_1, \ldots, z_{n^2})$ is the *same* as over $K(z_1, \ldots, z_r)$, hence has degree q. Apply Lemma 5 with $H_1, \ldots, H_{h+1}$ equal to the $n^2 + 1$ homogeneous polynomials $(\det(Y_{ij}))^p$, $z_1^n, \ldots, z_{n^2}^n$ of degree np in the n^2 indeterminates $Y_{11}, \ldots, Y_{nn}$. As $z_1, \ldots, z_{n^2}$ are algebraically independent over K, the polynomial $\det(Y_{ij})^p$ must be present in the corresponding relation (9), hence

$$q \leqslant M'' = pL = pn^2(pn+1)^{n^2-1}, \tag{10}$$

since (after division by the coefficient in (9) of the highest power of $\det(Y_{ij})^p$) we obtain from (9) an equation for $\det(Y_{ij})$ with coefficients in $K(z_1, \ldots, z_{n^2})$ which must be a multiple of the minimal equation.

It remains to prove Lemma 5. If we annihilate in a relation (9) the coefficient of each monomial in $u_1, \ldots, u_h$, we obtain a system of linear homogeneous equations between the $(L+1) \cdots (L+h)/h!$ unknowns $C_{\alpha_1\alpha_2\ldots\alpha_{h+1}}$. The number of these equations is

$$(mL+1) \cdots (mL+h-1)/(h-1)!.$$

Hence there will be nontrivial solutions to this system provided $(L+1)\cdots(L+h) > h(mL+1)\cdots(mL+h-1)$ and *a fortiori* this will be the case if

$$(L+1)^h > h(mL+h-1)^{h-1}; \tag{11}$$

but since the relation $L \geqslant h(m+1)^{h-1}$ implies $L \geqslant h$, it also implies $mL + h - 1 < (m+1)L$, hence

$$L+1 > h(m+1)^{h-1} > h\left(\frac{mL+h-1}{L+1}\right)^{h-1}$$

which is (11).

Appendix: A Short Digest of Non-Commutative Algebra

The purpose of this appendix is to give self-contained proofs of Maschke's Theorem and the Schur Commutation Theorem. The definitions of modules, algebras, etc. can all be found in "Fundamental Concepts of Algebra" by Chevalley. We shall be concerned exclusively with the following categories:

(i) The category $\mathscr{V}$ of *finite dimensional* Ω-vector spaces over the algebraically closed field Ω of *characteristic zero* whose morphisms are linear maps.

(ii) The category $\mathfrak{A}$ of *finite dimensional* Ω-algebras A with identity 1 whose morphisms are algebra homomorphisms φ such that $\varphi(1) = 1$. We consider Ω as a subset of A by the imbedding $\lambda \to \lambda \cdot 1$ and hence Ω is in the center of A.

(iii) The category $\mathscr{M}$ whose objects E are the objects of $\mathscr{V}$ which are left A-modules, where A is an object of $\mathfrak{A}$, and whose morphisms are A-module homomorphisms. We require that an A-module E be unitary, that is, $1 \cdot x = x$ for all $x \in E$.

Part I. Simple Modules

Suppose E is an A-module in $\mathscr{M}$. By an endomorphism of E is meant an element of $\mathrm{Hom}_\Omega(E, E) = \mathrm{End}(E)$. If $s \in A$, then $x \to sx$ defines an element $\underline{U}(s)$ of $\mathrm{End}(E)$ which we consider as a matrix. The mapping $s \to \underline{U}(s) : A \to \mathrm{End}(E)$ is an algebra homomorphism, and thus to any A-module E is associated a representation $A \to \mathrm{End}(E)$.

Recall that E and F are isomorphic A-modules if there exists a vector space isomorphism $\varphi : E \to F$ such that $\varphi(sx) = s\varphi(x)$ for all $s \in A$ and $x \in E$. If $\underline{U} : A \to \mathrm{End}(E)$ and $\underline{V} : A \to \mathrm{End}(F)$ are the corresponding representations, and if $\underline{P}$ is the matrix representing φ, then $\underline{P}\,\underline{U}(s) = \underline{V}(s)\,\underline{P}$. Thus $\underline{P}\,\underline{U}(s)\,\underline{P}^{-1} = \underline{V}(s)$, so $\underline{U}$ and $\underline{V}$ are *equivalent* representations. More generally, if $\varphi : E \to F$ is any homomorphism of A-modules, then $\underline{P}\,\underline{U}(s) = \underline{V}(s)\,\underline{P}$ for $s \in A$.

We have thus seen that an A-module E can be viewed as a representation of A in $\mathrm{End}(E)$ and conversely. If H is a linear subspace of E such that for $x \in H$ and $s \in A$ we have $sx \in H$, then H is called a *submodule* of E. Equivalently, a linear subspace H of E stable under the (image of the) representation of A in $\mathrm{End}(E)$ is a submodule of E.

Definition. An A-module $E \neq \{0\}$ is said to be *simple* if E has no proper nontrivial submodule. The representation corresponding to a simple A-module is called an *irreducible representation* of A.

If $H \neq \{0\}$ is a subspace of E, we may always extend a basis $e_1, \ldots, e_k$ of H to a basis $e_1, \ldots, e_n$ of E. The condition $\underline{U}(s)\, x \in H$ if $x \in H$ is that for some nonsingular $n \times n$ matrix $\underline{P}$,

$$\underline{P}^{-1}\underline{U}(s)\,\underline{P} = \begin{pmatrix} \underline{V}(s) & \underline{R}(s) \\ 0 & \underline{W}(s) \end{pmatrix} \tag{1}$$

where $\underline{V}(s)$ is a $k \times k$ matrix. The condition that a representation of A in $\mathrm{End}(E)$ be irreducible is that there is no $\underline{P}$ such that (1) holds for all $s \in A$.

Proposition. *An A-module $E \neq \{0\}$ is simple if and only if for each nonzero $x \in E$ we have $A \cdot x = E$.*

For $A \cdot x$ is a nontrivial submodule of E if $x \neq 0$.

Example. If $E \neq \{0\}$ is a vector space over Ω, then by the above proposition, E is a simple $\mathrm{End}(E)$-module.

We are now ready to state our first fundamental result.

Schur's Lemma I. *Suppose E is a simple A-module, F is an arbitrary A-module, and $u : E \to F$ is a homomorphism of A-modules. Then either $u = 0$ or u is injective. If F is also simple, then $u = 0$ or u is bijective.*

Proof. The first assertion follows since if $u \neq 0$ then $u^{-1}(0)$ is a proper submodule of E and hence $\{0\}$. If F is also simple and $u \neq 0$, then $u : E \to u(E)$ is an isomorphism. As $u(E)$ is a nonzero submodule of F, it must equal F.

Remark. In terms of representations, Schur's Lemma I is the following: If $s \to \underline{U}(s)$ and $s \to \underline{V}(s)$ are two irreducible representations of A, and if there exists a matrix $\underline{P}$ such that $\underline{P}\,\underline{U}(s) = \underline{V}(s)\,\underline{P}$, then either $\underline{P} = \underline{O}$ or $\underline{P}$ is invertible.

For an A-module E, let $\mathrm{End}_A(E) = \mathrm{Hom}_A(E, E)$.

Schur's Lemma II. *Let E be a simple A-module. Then* $\mathrm{End}_A(E) \simeq \Omega$.

Proof. By Schur's Lemma I, $\mathrm{End}_A(E)$ is a division ring or *sfield*. Let $\mu \in \mathrm{End}_A(E) - \Omega$, and let $\Omega(\mu)$ be the smallest subsfield of $\mathrm{End}_A(E)$ containing Ω and μ. Since Ω is in the center of $\mathrm{End}_A(E)$, $\Omega(\mu)$ is a field.

Since $\mathrm{End}_A(E) \subset \mathrm{End}(E)$, and since $\mathrm{End}(E)$ is finite dimensional over Ω, the powers $1, \mu, \mu^2, \ldots, \mu^n$ must be linearly dependent for some n. This implies μ is a root of a polynomial over Ω, and it follows that $\Omega(\mu)$ is a finite extension of Ω. But Ω, being algebraically closed, has no proper finite extensions, so $\Omega(\mu) = \Omega$, and the lemma follows.

Thus, if $\underline{U}$ is an irreducible representation of A and $\underline{P}\underline{U}(s) = \underline{U}(s)\,\underline{P}$ for all $s \in A$, then $\underline{P} = \lambda \underline{I}$ for some $\lambda \in \Omega$.

Part II. Semi-Simple Modules

Let us recall some elementary concepts from the theory of modules. If M is a module and $(N_i)_{i\in I}$ is a family of submodules of M, then $\Sigma_{i\in I} N_i$ denotes the submodule N of M generated by the set $\cup\{N_i : i \in I\}$. N is called the *sum* of the family $(N_i)_{i\in I}$. If, for every $i \in I$, $N_i \cap \Sigma_{j\neq i} N_j = \{0\}$, we say that N is the *direct sum* of the family $(N_i)_{i\in I}$ and write $N = \Sigma_{i\in I} N_i$ (direct). If $I = \{1,\ldots, p\}$, $\Sigma_{i\in I} N_i$ (direct) is often written $N_1 \oplus N_2 \oplus \cdots \oplus N_p$.

Theorem. *For an A-module $M \neq \{0\}$ in $\mathcal{M}$ the following are equivalent to each other*:

(i) *M is a (possibly infinite) sum of simple submodules*;

(ii) *M is a direct sum of finitely many simple submodules*;

(iii) *every submodule P of M is a direct summand.*

(*The exchange property.*) *If M is the sum of a family of simple submodules, say $M = \Sigma_{\alpha\in I} N_\alpha$, and if P is a submodule of M, there exists a finite subset $\{\alpha_1, \ldots, \alpha_k\}$ of I such that $M = P \oplus N_{\alpha_1} \oplus \cdots \oplus N_{\alpha_k}$.*

The proof is found in Chevalley, page 61.

Definition. A module M satisfying any of the above conditions is called *semi-simple.*

Corollary. *Quotients and submodules of semi-simple modules are semi-simple.*

Example. Every finite dimensional Ω-vector space is a semi-simple Ω-module.

Definition. A semi-simple module M is called *isotypic* of type N if and only if $M = N_1 \oplus \cdots \oplus N_k$, where each N_i is simple, and for any i, N_i is isomorphic to N.

Proposition. *A semi-simple A-module M can be expressed uniquely as* $M_1 \oplus \cdots \oplus M_r$, *where, for each i,* M_i *is isotypic, and, for* $i \neq j$, *the types of* M_i *and* M_j *are distinct. Every simple submodule* $P \subset M$ *is contained in exactly one* M_i, *and, in that case, P is isomorphic to the type of* M_i. *Finally, if* $i \neq j$, M_i *and* M_j *are not isomorphic as A-modules.*

Proof. Express $M = N_1 \oplus \cdots \oplus N_p$ with each N_i simple. Set $M_1 = \sum_{N_j \cong N_1} N_j$ and let i_2 be the smallest index such that N_{i_2} is not isomorphic to N_1. Set

$$M_2 = \sum_{N_j \cong N_{i_2}} N_j .$$

If we proceed in this way until exhausting all the N_i we obtain a direct sum decomposition $M = M_1 \oplus M_2 \oplus \cdots \oplus M_k$ where each M_i is isotypic and M_i and M_j have different types for $i \neq j$. If N is any simple submodule, $N \cong N_i$ for some i, for the contrary would imply, by Schur's Lemma I, that each projection $p_i : M \to N_i$ restricted to N is 0. Furthermore, $p_j : N \to N_j$ is 0 if N_j is not isomorphic to N_i. Hence N is in some M_k. Finally, if $M_i \cong M_j$, they have isomorphic simple submodules, so $i = j$.

Definition. The decomposition $M = M_1 \oplus \cdots \oplus M_r$ of the previous proposition is called the *isotypic decomposition* of M, and the M_i are called the *isotypic components* of M.

We conclude this section by computing $\mathrm{End}_A(M)$ for a semi-simple A-module M. We will view $\mathrm{End}_A(M)$ as an Ω-algebra and hence all isomorphisms will be isomorphisms of algebras. Let $\underline{M}_n(\Omega)$ denote the algebra of $n \times n$ matrices over Ω. Let $M = M_1 \oplus \cdots \oplus M_r$ be the isotypic decomposition of M and write $M_i = N_{i1} \oplus \cdots \oplus N_{ij_i}$, where each N_{ik} is simple and isomorphic to N_i, the type of M_i.

If $\mu \in \mathrm{End}_A(M)$, then $\mu \mid N_{ik}$ is 0 or injective, and in the latter case $\mu(N_{ik}) \cong N_{ik}$. Hence $\mu(N_{ik}) \subset M_i$ for each $k = 1,\ldots, j_i$, and so $\mu(M_i) \subset M_i$. Thus $\mathrm{End}_A(M) \cong \mathrm{End}_A(M_1) \oplus \cdots \oplus \mathrm{End}_A(M_r)$. Furthermore, $\mathrm{End}_A(M_i) \cdot \mathrm{End}_A(M_j) = \{0\}$ if $i \neq j$. We express these two facts by writing $\mathrm{End}_A(M) = \prod_{i=1}^r \mathrm{End}_A(M_i)$.

We shall now determine $\mathrm{End}_A(M)$ for M *isotypic*. Suppose $M = N_1 \oplus \cdots \oplus N_s$ where, for any i, N_i is simple and $N_i \cong N$. For $j = 1,\ldots,s$, let $p_j : M \to N_j$ be the projection of M onto N_j. The p_j allow one to associate to any $u \in \mathrm{End}_A(M)$ an $s \times s$ matrix $\underline{U} = (u_{ij})$, where $u_{ij} \in \mathrm{Hom}_A(N_j, N_i)$ for each i and j. Specifically,

$$u_{ij} = p_i \circ (u \mid N_j) : N_j \to N_i .$$

It is easy to see that for $u, v \in \text{End}_A(M)$ and for $w = vu$, we have $\underline{W} = \underline{V}\,\underline{U}$. In other words, $w_{ij} = \sum_{k=1}^{s} v_{ik}u_{kj}$.

Now choose a fixed isomorphism $\varphi_i : N_i \xrightarrow{\sim} N$ for every i. Then the map $\psi_{ij} : \text{Hom}_A(N_j, N_i) \to \text{End}_A(N)$ given by $\psi_{ij}(f) = \varphi_i f \varphi_j^{-1}$ is an isomorphism of Ω-modules. Furthermore, if $f_{kj} \in \text{Hom}_A(N_j, N_k)$ and $g_{ik} \in \text{Hom}_A(N_k, N_i)$, then $\psi_{ij}(g_{ik} f_{kj}) = \psi_{ik}(g_{ik})\,\psi_{kj}(f_{kj})$. By Schur's Lemma II, $\text{End}_A(N) \cong \Omega$. (In fact, we proved that if $f \in \text{End}_A(N)$, then for some $\lambda \in \Omega$, $f(x) = \lambda x$ for any $x \in N$). These facts imply that the composition

$$\Psi : u \to (u_{ij}) \to (\psi_{ij}u_{ij}) : \text{End}_A(M) \to \underline{M}_s(\Omega)$$

defines an Ω-algebra homomorphism. It is immediate that Ψ is bijective for Ψ^{-1} can easily be defined. Combining everything we have shown, we get the

Theorem. *If M is a semi-simple A-module with isotypic decomposition $M = M_1 \oplus \cdots \oplus M_r$, then $\text{End}_A(M) \cong \prod_{i=1}^{r} \underline{M}_{j_i}(\Omega)$, where $j_i = \dim_A M_i$ for each i.*

Note that $\dim_A M = j$ means that one can write $M = N_1 \oplus \cdots \oplus N_j$, each N_i being a simple A-module.

Remark. In terms of representations, semi-simple modules correspond to *completely reducible* representations $\underline{U}$: i.e.,

$$\underline{U}(s) = \begin{pmatrix} \underline{U}_1(s) & & & 0 \\ & \underline{U}_2(s) & & \\ & & \ddots & \\ 0 & & & \underline{U}_p(s) \end{pmatrix}$$

with $\underline{U}_i(s)$ irreducible.

Part III. Semi-Simple Algebras

Theorem. *For an Ω-algebra A the following are equivalent:*

(i) *every A-module is semi-simple;*

(ii) *A_s (A viewed as a left A-module) is semi-simple.*

Proof. (i) trivially implies (ii). Suppose then that A_s is semi-simple and let M be an A-module. Clearly $M = \sum_{m \in M} A \cdot m$. There is a natural surjection $A_s \to A \cdot m$ defined by $a \to am$. This is a homomorphism of (left) A-modules, and hence $A \cdot m = A_s/P$, for some submodule P of A_s. But A_s/P being a quotient of a semi-simple module is itself semi-simple. Thus it follows that M is a sum of simple submodules, so M is semi-simple.

Definition. An algebra A is said to be *semi-simple* if either (i) or (ii) of the above holds. If A is semi-simple, a simple submodule of A_s is called a *minimal left ideal* of A.

Example. Let $\underline{M}_n(\Omega)$ denote the algebra of $n \times n$ matrices over Ω, and let $\underline{E}_{ij} \in \underline{M}_n(\Omega)$ denote the matrix with ij-th entry 1 and zeros elsewhere. Then $\mathscr{L}_i = \Omega\underline{E}_{1i} \oplus \cdots \oplus \Omega\underline{E}_{ni}$ (i-th column) is a minimal left ideal in $\underline{M}_n(\Omega)$, and $\underline{M}_n(\Omega) = \mathscr{L}_1 \oplus \cdots \oplus \mathscr{L}_n$. It follows that $\underline{M}_n(\Omega)$ is semi-simple.

There is an elegant description of the minimal left ideals of $\underline{M}_n(\Omega)$. These correspond exactly to the hyperplanes of $E = \Omega^n$ in the following way. Choose a basis $e_1, \ldots, e_n$ of E and identify $\mathrm{End}(E)$ with $A = \underline{M}_n(\Omega)$ in the usual way. If H is a hyperplane in E, let

$$\mathscr{L}_H = \{u \in \mathrm{End}(E) : \ker u \supset H\}.$$

Then $\mathscr{L}_H$ is a minimal left ideal in A, and every minimal left ideal in A is an $\mathscr{L}_H$. To see that $\mathscr{L}_H$ is minimal, choose an isomorphism $\varphi: H \to H_1$, where H_1 is the hyperplane spanned by $e_2, \ldots, e_n$. Define

$$\varphi^* \in \mathrm{Hom}_A(\mathscr{L}_H, \mathscr{L}_1)$$

by $\varphi^* u = u\varphi$; φ^* is clearly an isomorphism so $\mathscr{L}_H$ is minimal.

If $\mathscr{L}$ is a minimal left ideal in A, let $u \neq 0 \in \mathscr{L}$, and let V be a subspace of $u(E)$ of codimension one. Then $H = u^{-1}(V)$ is a hyperplane in E, and we assert that $\mathscr{L}_H = \mathscr{L}$. Choose $v \in A$ such that $v(V) = \{0\}$ but $v(u(E)) \neq \{0\}$. Then $vu \neq 0$, $vu \in \mathscr{L}_H$, and hence $\mathscr{L}_H = A(vu) = \mathscr{L}$.

The minimal right ideals of A admit a similar classification. Each of these consists of all $u \in A$ such that $\mathrm{Im}\, u$ is contained in a fixed line in E.

Molien's Theorem. *Every semi-simple algebra A is isomorphic to a direct product of matrix rings over Ω; that is, $A \cong \prod_{j=1}^{r} \underline{M}_{n_j}(\Omega)$, where r is the number of isotypic components of A.*

Proof. We first need a general lemma.

Lemma. *Let R be a ring with identity* 1. *Then* $\mathrm{End}_R(R_s) \cong R^0$, *the opposite ring of R.*

Proof. Recall that R^0 is the ring R with multiplication defined in reverse order. For $x \in R$ define $t_x \in \mathrm{End}_R(R_s)$ by $t_x(y) = yx$. Since $t_{xx'} = t_{x'}t_x$, t is a (ring) homomorphism of R^0 into $\mathrm{End}_R(R_s)$. If $u \in \mathrm{End}_R(R_s)$ and $u(1) = x$, then $u(y) = yx$ for all $y \in R$ and $t_x = u$. If $t_x = 0$, then $t_x(1) = 1 \cdot x = 0$, so $x = 0$. Thus t is a bijection.

If A is an Ω-algebra, then the isomorphism $A^0 \cong \mathrm{End}_A(A_s)$ is an isomorphism of algebras. Hence by an earlier theorem, if A is semi-simple, then $A^0 \cong \prod_{j=1}^r \underline{M}_{n_j}(\Omega)$. Since, however, Ω is a field, the mapping $\underline{X} \to {}^t\underline{X}$ defines an isomorphism $\underline{M}_{n_j}(\Omega) \cong \underline{M}_{n_j}(\Omega)^0$, so Molien's Theorem is proved.

Note that by Part II, each $\underline{M}_{n_j}(\Omega)$ corresponds to an isotypic component of A under the isomorphism. Thus if $A_1, \ldots, A_r$ are the isotypic components of A, $A = \prod_{i=1} A_i$.

Proposition. *Let A be a semi-simple algebra, and let M be an A-module. Let $A_1, \ldots, A_r$ be the isotypic components of A_s. Then $A_1M, \ldots, A_rM$ are precisely the isotypic components of M.*

Proof. Obviously $M = \Sigma_i A_i M$. The sum $\Sigma A_i M$ is direct, for if

$$x \in A_jM \cap \Sigma_{i \neq j} A_i M,$$

and if e_j is the component of the unit of A in A_j, then $e_j x = x$ and $e_i \Sigma_{i \neq j} A_j M = \{0\}$. Thus $x = 0$.

If $N \subset M$ is simple, then from $N = A_1N \oplus \cdots \oplus A_rN$, we must have $N = A_iN$ for exactly one i. Hence $N \subset A_iM$. Two isomorphic simple submodules of M must be contained in the same A_iM, so $A_1M, \ldots, A_rM$ are the isotypic components of M.

Theorem (Molien). *For a semi-simple algebra A over Ω, the following are equivalent:*

(i) *there is only one type of simple A-module;*

(ii) *there is no nontrivial proper two-sided ideal in A;*

(iii) $A \cong \underline{M}_n(\Omega)$.

Proof. That (i) is equivalent to (iii) follows from Molien's theorem and the fact that the number of types of simple submodules is the number of isotypic components. That (ii) implies (i) follows since each isotypic component is a two-sided ideal in A. We have only to show that (iii) implies (ii). Let $\mathfrak{A} \neq \{0\}$ be a two sided ideal in $A = \underline{M}_n(\Omega)$ and let $\mathscr{L}_H$ be a minimal left ideal in $\mathfrak{A}$ corresponding to the hyperplane H of Ω^n. Recall that $\mathscr{L}_H = \{\underline{U} \in \underline{M}_n(\Omega) : \ker \underline{U} \supset H\}$. Now if H' is any other hyperplane of Ω^n and s is an isomorphism of Ω^n mapping H' onto H, then $\mathscr{L}_{H'} = \mathscr{L}_H s$. Since $\mathfrak{A}$ is a right ideal $\mathscr{L}_{H'} \subset \mathfrak{A}$. This shows that every minimal left ideal in A is contained in $\mathfrak{A}$, so $\mathfrak{A} = A$ since A is semi-simple (and thus is a direct sum of minimal left ideals).

Definition. A semi-simple algebra A is said to be *simple* if any of (i), (ii), or (iii) holds for A.

Thus any semi-simple algebra A decomposes as $A = \prod_{i=1}^{r} A_i$, where each A_i is simple. This decomposition of A is called the *simple decomposition* of A, and the A_i are called the *simple components* of A. Furthermore, any minimal two sided ideal B of A is one of the A_i. For

$$B = A_1B \oplus \cdots \oplus A_rB,$$

and since each A_iB is a two sided ideal in A, the minimality of B implies $B = A_jB$ for some j. But $A_jB \subset A_j$, so $B = A_j$, by the simplicity of A_j. Thus we have the

Proposition. *Let A be a semi-simple algebra with simple components $A_1,\ldots, A_r$. Then every two-sided ideal in A is of the form $\prod_{j\in F} A_j$, where F is any finite subset of $\{1, 2,\ldots, r\}$. Therefore any quotient of A by a two-sided ideal in A is again semi-simple. In particular, the homomorphic image of a semi-simple algebra is semi-simple.*

PART IV. MASCHKE'S THEOREM AND SCHUR'S COMMUTATION THEOREM

Suppose G is a finite group, and let $\Omega[G]$ be the group algebra of G. $\Omega[G]$ can be described briefly as follows. As an Ω-vector space, $\Omega[G]$ is the vector space whose basis is the set of elements of G. $\Omega[G]$ becomes an Ω-algebra by defining

$$(\Sigma_{r\in G}\lambda_r r) \cdot (\Sigma_{s\in G}\mu_s s) = \Sigma_{r,s\in G}(\lambda_r\mu_s)\, r \cdot s,$$

where $\lambda_r, \mu_s \in \Omega$. The unit $e \in G$ is the identity of $\Omega[G]$.

Maschke's Theorem. *If G is a finite group of order m, then the group algebra $\Omega[G]$ is semi-simple.*

Proof. Let A denote $\Omega[G]$, and suppose E is any finite dimensional A-module. To show E is semi-simple we will prove that any submodule F of E has a supplementary submodule W. Let W be a vector space supplement of F in E; that is, $E = W \oplus F$ as a vector space. Then there exists a projection $p : E \to F$. Define $p_0 : E \to F$ by

$$p_0(x) = m^{-1}\Sigma_{s\in G} sp(s^{-1}x).$$

p_0 is clearly linear. If $t \in G$,

$$\begin{aligned} mp_0(tx) &= \Sigma_{s\in G} sp(s^{-1}tx) \\ &= \Sigma_{s\in G} t(s^{-1}t)^{-1}\, p(s^{-1}tx) \\ &= \Sigma_{u\in G} tup(u^{-1}x) \\ &= t(mp_0(x)). \end{aligned}$$

Thus p_0 is an A-homomorphism. For $y \in F$, $p(s^{-1}y) = s^{-1}y$, hence $p_0(y) = y$, and we have $p_0 p_0 = p_0$. Therefore $(1 - p_0)\,E$ is an A-submodule of E supplementary to F.

Maschke's result has an elegant generalization first noticed by Hurwitz in 1897.

Theorem. *Let G be a compact group and let ds be (normalized) Haar measure on G. Let $s \to \underline{U}(s)$ be a continuous representation of G in* $\mathbf{GL}(E)$, *where E is a finite dimensional complex vector space. If $F \subset E$ is a G-module (that is, F is stable under $\underline{U}(G)$), then F has a supplementary G-module H.*

The proof is exactly the same as the proof of Maschke's Theorem. One just replaces $m^{-1}\Sigma_{s\in G}$ by $\int_G ds$.

Corollary. *Every compact subgroup of* $\mathbf{GL}(n; \mathbf{C})$ *is reductive.*

There is a convenient way of determining the number of simple components of a semi-simple algebra A which will be applied to obtain an expression for the number of simple components of $\Omega[\mathscr{G}_f]$. Let A be semi-simple, and let $A = \prod_{i=1}^r A_i$ be the decomposition of A in simple components. The center Z of A decomposes correspondingly as $Z = \prod_{i=1}^r Z_i$, where Z_i is the center of A_i. But $A_i \simeq \underline{M}_n(\Omega)$ for some n, and since the center of $\underline{M}_n(\Omega)$ is just Ω, we have the

Proposition. *The number of simple components of a semi-simple algebra A is equal to the dimension over Ω of the center of A.*

Proposition. *If G is any finite group, then the number of simple components of $\Omega[G]$ is precisely the number of conjugacy classes in G.*

Proof. By the last proposition, we need to compute $\dim_\Omega Z$, where Z is the center of $\Omega[G]$. Let $z = \Sigma_{s\in G}\lambda_s s$ be an element of the center. Then the condition $tzt^{-1} = z$ for each $t \in G$ implies that $\Sigma_{s\in G}\lambda_s tst^{-1} = \Sigma_{s\in G}\lambda_s s$. Hence if s and s' are conjugate, then $\lambda_{s'} = \lambda_s$. Let $\gamma_1, \ldots, \gamma_r$ be the conjugacy classes of G. Then z can be expressed uniquely as

$$\lambda_1 \Sigma_{s\in\gamma_1} s + \cdots + \lambda_r \Sigma_{s\in\gamma_r} s,$$

where $\lambda_1, \ldots, \lambda_r \in \Omega$. Since every element of this form is in the center, we have the result.

We will now prove Schur's Commutation Theorem whose statement we postpone until the end of the proof. Suppose that B is a semi-simple subalgebra of $A = \underline{M}_N(\Omega)$ (hence the unit of A lies in B). We wish to determine how B is situated in A and how the *commutant*

$$C = \{\underline{W} \in \underline{M}_N(\Omega) : \underline{W}\underline{U} = \underline{U}\underline{W} \quad \text{for all} \quad \underline{U} \in B\}$$

of B is situated with respect to B.

Let $V = \Omega^N$. Thus $A = \operatorname{End}(V)$, and so we regard B as a subalgebra of $\operatorname{End}(V)$. Since $1 \in B$, V is a B-module, and since B is semi-simple, V is semi-simple also. Let $B_1, \ldots, B_r$ be the simple components of B. Then $V = B_1V \oplus \cdots \oplus B_rV$ is the isotypic decomposition of V.

Each B_iV can be written as a direct sum $V_{i1} \oplus \cdots \oplus V_{im_i}$ of simple submodules of V such that, for each $i = 1, \ldots, r$ and $j = 1, \ldots, m_i$, there exists a B-module isomorphism $\varphi_{ij} : V_{i1} \to V_{ij}$. We take φ_{i1} to be the identity. Let $e_i^{(1)}, \ldots, e_i^{(n_i)}$ be a basis of V_{i1} for each i. Then

$$\varphi_{ij}(e_i^{(1)}), \ldots, \varphi_{ij}(e_i^{(n_i)})$$

is a basis of V_{ij} if $1 \leqslant j \leqslant m_i$. These bases for the V_{ij} combine to give a basis

$$e_1^{(1)}, \ldots, e_1^{(n_1)}, \varphi_{12}(e_1^{(1)}), \ldots, e_2^{(1)}, \ldots, e_2^{(n_2)}, \varphi_{22}(e_2^{(1)}), \ldots$$

of V, and, with respect to this basis, an endomorphism $\underline{U}$ of B has matrix

$$\begin{bmatrix} \underline{U}_{11} & & & & & & 0 \\ & \ddots & & & & & \\ & & \underline{U}_{1m_1} & & & & \\ & & & \underline{U}_{21} & & & \\ & & & & \ddots & & \\ 0 & & & & & & \underline{U}_{rm_r} \end{bmatrix}, \tag{2}$$

with $\underline{U}_{ij} \in \underline{M}_{n_i}(\Omega)$. This follows from the fact that V_{ij} is a B-module. One should note that we have now replaced the subalgebra B by the semi-simple subalgebra $\underline{P}^{-1}B\underline{P}$ of $\underline{M}_N(\Omega)$ for some invertible $\underline{P} \in \underline{M}_N(\Omega)$.

The fact that $\varphi_{ij}(\underline{U}x) = \underline{U}\varphi_{ij}(x)$ for each $x \in V$ implies that $\underline{U}_{i1} = \cdots = \underline{U}_{im_i}$ for any i. Hence $\underline{U}$ has the form

$$\underline{U} = \begin{pmatrix} \underline{T}_1 & & 0 \\ & \ddots & \\ 0 & & \underline{T}_r \end{pmatrix} \tag{3}$$

where

$$\underline{T}_i = \begin{pmatrix} \underline{U}_i & & 0 \\ & \ddots & \\ 0 & & \underline{U}_i \end{pmatrix}. \tag{4}$$

The set of all matrices of the form (4) with m_i copies of $\underline{U} \in \underline{M}_{n_i}(\Omega)$ along the diagonal will be denoted by $\underline{M}_{n_i}^{m_i}(\Omega)$.

Each $\underline{U}_i$ is arbitrary in $\underline{M}_{n_i}(\Omega)$. To see this one first notes that V_{i1} is also a simple B_i-module. But since B_i is simple, $B_i \cong \underline{M}_{q_i}(\Omega)$ for some q_i. Thus Ω^{q_i} and V_{i1} are simple B_i-modules, hence isomorphic, so $q_i = n_i$. It follows that $B \cong \prod_{i=1}^{r} \underline{M}_{n_i}^{m_i}(\Omega)$. We have thus determined every semi-simple subalgebra of $\underline{M}_N(\Omega)$. Each such is determined by a system of positive integers $(m_1, \ldots, m_r, n_1, \ldots, n_r)$ with $\sum_{i=1}^{r} m_i n_i = N$.

We will now determine the commutant C of B. According to the expression (3) for $\underline{U} \in B$, write $\underline{W} \in C$ as

$$\begin{pmatrix} \underline{W}_{11} & \underline{W}_{12} & \cdots & \\ \underline{W}_{21} & \underline{W}_{22} & & \\ \vdots & & \ddots & \\ & & & \underline{W}_{rr} \end{pmatrix}$$

with W_{ij} a $p_i \times p_j$ matrix, where $p_i = m_i n_i$. The equation $\underline{U}\underline{W} = \underline{W}\underline{U}$ implies that $\underline{T}_i \underline{W}_{ij} = \underline{W}_{ij} \underline{T}_j$ for every $i, j = 1,..., r$. But this means $\underline{W}_{ij} = 0$ if $i \neq j$, since $\underline{U}_i$ and $\underline{U}_j$ are completely arbitrary. Thus $\underline{W} \in C$ is of the form

$$\begin{pmatrix} \underline{W}_1 & & 0 \\ & \ddots & \\ 0 & & \underline{W}_r \end{pmatrix}$$

with $\underline{W}_i \underline{T}_i = \underline{T}_i \underline{W}_i$ for all $T_i \in B_i$. We infer that $C = \prod_{i=1}^{r} C_i$, where C_i is the commutant of B_i.

Hence it suffices to determine the commutant C of a simple subalgebra B of $\underline{M}_Q(\Omega)$. As a B-module, $V = \Omega^Q$ is isotypic, so $V = V_1 \oplus \cdots \oplus V_m$, where the V_i are simple and have the same type. As above, we may choose a basis $f_i^{(1)},..., f_i^{(n)}$ for each V_i so that $\underline{U} \in B$ is a matrix of the form

$$\begin{pmatrix} \underline{S} & & 0 \\ & \ddots & \\ 0 & & \underline{S} \end{pmatrix} \qquad (m \text{ copies of } \underline{S}),$$

where $\underline{S} \in \underline{M}_n(\Omega)$; that is, $B \cong \underline{M}_m^n(\Omega)$.

If $\underline{W} \in C$, write

$$\underline{W} = \begin{pmatrix} \underline{W}_{11} & \cdots & \underline{W}_{1m} \\ \vdots & & \vdots \\ \underline{W}_{m1} & \cdots & \underline{W}_{mm} \end{pmatrix}$$

where $\underline{W}_{ki} \in \underline{M}_n(\Omega)$. The condition $\underline{W} \in C$ means that $\underline{S}\underline{W}_{ki} = \underline{W}_{ki}\underline{S}$ for every $\underline{S} \in \underline{M}_n(\Omega)$. But the center of $\underline{M}_n(\Omega)$ is precisely the set of matrices of form $\lambda \underline{I}_n$ for some $\lambda \in \Omega$. Hence, for any i and k, $\underline{W}_{ki} = \lambda_{ki} \underline{I}_n$ for some $\lambda_{ki} \in \Omega$. Thus,

$$\underline{W}(f_i^{(j)}) = \sum_{k=1}^{m} \lambda_{ki} f_k^{(j)}$$

if $\underline{W} \in C$. This means that if the basis is reordered as $f_1^{(1)}, f_2^{(1)},..., f_m^{(1)}, f_1^{(2)},...$, the matrix of $\underline{W}$ becomes

$$\begin{pmatrix} (\lambda_{ij}) & & 0 \\ & \ddots & \\ 0 & & (\lambda_{ij}) \end{pmatrix} \in M_m^n(\Omega),$$

with the $(\lambda_{ij}) \in \underline{M}_m(\Omega)$ completely arbitrary. Therefore we have proven

Schur's Commutation Theorem. *Let B be a semi-simple subalgebra of $\underline{M}_N(\Omega)$, and let $B_1, \ldots, B_r$ be the simple components of B. Let n_i be the dimension over Ω of any simple B_i-module. Then:*

(i) $B \cong \prod_{i=1}^r \underline{M}_{n_i}^{m_i}(\Omega)$, *where* $\underline{M}_{n_i}^{m_i}(\Omega)$ *is defined by* (4);

(ii) *if C denotes the commutant of B, then $C \cong \prod_{i=1}^r \underline{M}_{m_i}^{n_i}(\Omega)$. In particular, C is semi-simple.*

Let B be a simple subalgebra of $\underline{M}_N(\Omega)$ with C its commutant. Then $B \cong \underline{M}_n^m(\Omega)$ and $C \cong \underline{M}_m^n(\Omega)$, so it follows that C is simple. Let V_C be Ω^N viewed as a C-module. V_C is isotypic, and the simple submodules of V_C are precisely the submodules of V_C of dimension m over Ω (for every minimal left ideal of C has dimension m). For $1 \leqslant i \leqslant n$, the matrix $\underline{E}_i = \underline{E}_{i,i} + \underline{E}_{n+i,n+i} + \cdots + \underline{E}_{(m-1)n+i,(m-1)n+i}$ generates a minimal left ideal of B, and, for any i, $\dim_\Omega \underline{E}_i V = m$. In fact,

$$\underline{E}_i V = \Omega f_1^{(i)} \oplus \Omega f_2^{(i)} \oplus \cdots \oplus \Omega f_m^{(i)}.$$

By the commutation property, $C\underline{E}_i V = \underline{E}_i C V \subset \underline{E}_i V$, so $\underline{E}_i V$ is a simple C-module. Furthermore, $V_C = \underline{E}_1 V \oplus \cdots \oplus \underline{E}_n V$.

If B is semi-simple, we may obtain the simple submodules of V_C by considering the simple components of B separately. Suppose $b, b' \in B$ generate the same minimal left ideal in B. Then $b' = u_1 b$, $b = u_2 b'$, and $x \to u_1 x : bV \to b'V$, $x \to u_2 x : b'V \to bV$ are C-homomorphisms (by the commutation property) which are inverse to each other. Therefore $bV \cong b'V$.

Proposition. *Every simple submodule of V_C is isomorphic to one of the form bV, where b generates a minimal left ideal in B. If b, b' generate the same minimal left ideal, then $bV \cong b'V$.*

Bibliography

1. G. Gurevich, "Algebraic theory of invariants," Noordhoff (1964).

2. D. Mumford, Geometric invariant theory. *Erg. der Math.* **34** (1965).

3. D. Hilbert, Über die Theorie der algebraischen Formen. *Math. Ann.* **36**, p. 473 (1890).

4. D. Hilbert, Über die vollen Invariantensysteme. *Math. Ann.* **42**, p. 313 (1893).

5. M. Nagata, "Lectures on Hilbert's fourteenth problem." Tata Institute (1963).

6. I. Schur, "Vorlesungen über Invarianttheorie." Springer (1968).

7. B. L. van der Waerden, Über die fundamentalen Identitäten der Invarianttheorie. *Math. Ann.* **95**, p. 706 (1925).

8. B. L. van der Waerden, Reihenentwickelungen und Überschiebungen in der Invarianttheorie insbesondere im quaternären Gebiet. *Math. Ann.* **113**, p. 14 (1936).

9. R. Weitzenböck, "Invarianttheorie." Noordhoff (1923).

10. H. Weyl, "The classical groups." Princeton University Press (1939).

11. O. Zariski and P. Samuel, "Commutative algebra." Van Nostrand (1958–60)

12. J. Fogarty, "Invariant theory." Benjamin (1969).